D0225569

MAXIMA AND MINIMA WITHOUT CALCULUS

By
IVAN NIVEN

THE DOLCIANI MATHEMATICAL EXPOSITIONS

Published by
THE MATHEMATICAL ASSOCIATION OF AMERICA

The Dolciani Mathematical Expositions

NUMBER SIX

MAXIMA AND MINIMA WITHOUT CALCULUS

By
IVAN NIVEN
University of Oregon

Published and distributed by
THE MATHEMATICAL ASSOCIATION OF AMERICA

The Mathematical Association of America (Incorporated)
Library of Congress Catalog Card Number 80-81045

Complete Set ISBN 0-88385-300-0
Vol. 6 ISBN 0-88385-306-X

Printed in the United States of America

Current printing (last digit):
10 9 8 7 6 5 4 3 2 1

The DOLCIANI MATHEMATICAL EXPOSITIONS series of the Mathematical Association of America was established through a generous gift to the Association from Mary P. Dolciani, Professor of Mathematics at Hunter College of the City University of New York. In making the gift, Professor Dolciani, herself an exceptionally talented and successful expositor of mathematics, had the purpose of furthering the ideal of excellence in mathematical exposition.

The Association, for its part, was delighted to accept the gracious gesture initiating the revolving fund for this series from one who has served the Association with distinction, both as a member of the Committee on Publications and as a member of the Board of Governors. It was with genuine pleasure that the Board chose to name the series in her honor.

The books in the series are selected for their lucid expository style and stimulating mathematical content. Typically, they contain an ample supply of exercises, many with accompanying solutions. They are intended to be sufficiently elementary for the undergraduate and even the mathematically inclined high-school student to understand and enjoy, but also to be interesting and sometimes challenging to the more advanced mathematician.

The following DOLCIANI MATHEMATICAL EXPOSITIONS have been published.

Volume 1: MATHEMATICAL GEMS, by Ross Honsberger

Volume 2: MATHEMATICAL GEMS II, by Ross Honsberger

Volume 3: MATHEMATICAL MORSELS, by Ross Honsberger

Volume 4: MATHEMATICAL PLUMS, edited by Ross Honsberger

Volume 5: GREAT MOMENTS IN MATHEMATICS (BEFORE 1650), by Howard Eves

Volume 6: MAXIMA AND MINIMA WITHOUT CALCULUS, by Ivan Niven

ILLUSTRATIONS BY DAVID E. LOGOTHETTI AND THE AUTHOR

TO BETTY

PREFACE

Our purpose is to bring together the principal elementary methods for solving problems in maxima and minima, except for two techniques that are treated adequately in standard textbooks. Calculus is deliberately omitted from our discussions, as are optimization processes through linear programming and game theory. In view of the many books and courses available on these subjects, our purpose is to complement these sources, not to compete with them. Thus there is a deliberate imbalance in this book, leaning toward methods in algebra and geometry that are not so widely known. Also, as the reader will readily note, our preference is often to solve geometric problems by reformulating them in an algebraic setting rather than by using purely geometric methods. Another author might do it another way, but, as the old saying goes, one man's fish is another man's poisson.

Calculus is such a systematically organized subject, providing as it does a step-by-step procedure for solving extremal problems, that its champions often regard alternative methods as trick procedures of limited usefulness. We attempt to counter this view by unifying these alternative procedures as much as possible. In this way the techniques can be perceived not simply as special devices of limited usefulness but as more general methods offering wider application. Thus we emphasize the line of argument that will accommodate many questions, rather than the brilliant shot that polishes off one and only one problem in splendid isolation.

Although calculus does provide a powerful and systematic technique for solving *some* problems in maxima and minima, the method is not universal. There are many questions that are awkward, if not impossible, by elementary calculus. Consider for example the question asked in calculus books of finding, among rectangles of a given perimeter, the one with largest area. The broader question of finding the *quadrilateral* of largest area among those of a given perimeter is not well suited to elementary calculus. Such questions are grist for our mill. Thus we follow a simple maxim: If a problem can be solved more simply by calculus, leave it to calculus.

Extremal questions are very close to problems in inequalities, so it is not surprising that this topic pops up quite regularly. However, our interest is not in inequalities per se, but only to the extent that they contribute to the solution of the extremal problems.

What background is needed to read this book? It is written for an audience at or near the maturity level of second- and third-year students in North American universities and colleges, assuming a good working knowledge of precalculus mathematics. Although calculus is not a prerequisite, a prior knowledge of that subject would enhance the reader's comprehension.

Although various techniques from geometry are introduced, there are three methods that are not used: orthogonal and other projections, vector analysis, and the geometry of complex numbers. These methods could have been used to simplify some of the solutions, but their introduction would have led us too far afield.

Chapter 1 contains some highlights of the background material needed, with the principal subject matter of the book starting in the second chapter. Although some readers will be able to proceed to Chapter 2 almost directly, Section 1.1 should be given some attention since it includes some basic agreements about language and notation.

The plan of the book is to proceed from easy problems to harder ones. For example, consider the isoperimetric problem in the plane: among all simple closed curves of a given length, which encloses the maximum area? This problem is solved in Chapter 4, in Section 4.3 to be specific, under the assumption that a solution exists. We return to the topic in Chapter 12, where the problem is solved without assuming the existence of a solution. From a logical standpoint these two chapters should be combined—in fact, with parts of Chapter 4 discarded because Chapter 12 is more general in its scope. However, the later chapter is not as easy to follow as the earlier one, which is much more elementary.

Chapters 2 to 6 are intended to be read in succession, each dependent on the earlier ones. These chapters are prerequisites for Chapters 7, 8 and 12, which can be read independently. Chapters 9, 10, and 11 also can be read independently, with Chapters 2 and 3 as needed background.

There are many problems for the reader scattered through the book. They are identified by a letter and a number; for example, E11 is the eleventh problem in Chapter 5. At the back of the book answers are given for all problems as needed, as well as solutions for most. The reader is urged, of course, to try the problems for herself or himself, turning to the solutions as a last resort. There are no exercises or drill problems, because the work is intended primarily as a resource book, not a textbook. The author has used parts of the material in the book, however, in an experimental course several times.

The notes at the ends of the chapters give not only sources of the material, but also suggestions for further reading. Although some references are listed in the body of the book, most are collected in one master list at the end, with the authors in alphabetical order. No attempt has been made to give a complete bibliography of the subject.

A first version of the manuscript was read by members of the committee on the Dolciani series, and by G. D. Chakerian, Basil Gordon, and Roy Ryden. I was very fortunate to get their constructive suggestions, which have resulted in extensive improvements. I am also grateful to many people for suggesting topics, problems, and references that might have been overlooked; especially I mention M. S. Klamkin, L. H. Lange, and the late C. D. Olds in this connection.

IVAN NIVEN

University of Oregon

CONTENTS

CHAPTER 1

BACKGROUND MATERIAL

This chapter contains the definitions, notations, conventions and background results needed for an understanding of the book. Although for many readers it will suffice to skim this chapter, the first section is somewhat crucial since it contains agreements about the use of language and notation. But the really substantive discussions of maxima and minima begin with Chapter 2, so the reader is urged to move on to that as quickly as possible.

1.1 Language and Notation. If a and b are any real numbers, the assertion that a is greater than b means that $a - b$ is positive, and this can be written in several equivalent forms:

$$a > b, \quad a - b > 0, \quad b < a, \quad b - a < 0.$$

Similarly, the statement that a is greater than or equal to b means that $a - b$ is positive or zero, and we can write

$$a \geq b, \quad a - b \geq 0, \quad b \leq a, \quad b - a \leq 0.$$

The notation $\max(a, b, c)$ denotes the largest, or the *maximum*, among the real numbers a, b, c. For example

$$\max(2, 3, 5) = 5, \quad \max(2, 3, -5) = 3, \quad \max(3, 3, -5) = 3.$$

In general, let $a_1, a_2, \ldots, a_n$ be any finite collection of real numbers, not necessarily all distinct. The equation

$$\max(a_1, a_2, \ldots, a_n) = a_j,$$

where j is an integer among $1, 2, 3, \ldots, n$, means that all the inequalities

$$a_j \geq a_1, \quad a_j \geq a_2, \quad a_j \geq a_3, \quad \ldots, a_j \geq a_n$$

hold. Similarly the *minimum* of a finite collection of real numbers is denoted by

$$\min(a_1, a_2, \ldots, a_n) = a_k,$$

and this means that all the inequalities

$$a_k \leq a_1, \quad a_k \leq a_2, \quad a_k \leq a_3, \quad \ldots, a_k \leq a_n$$

hold.

For infinite sets of real numbers, there may or may not be a maximum or a minimum. As a simple example, there is no smallest positive real number because if r is any positive number, $r/2$ is smaller. If a and b are any given real numbers with $a < b$, the set of numbers x satisfying

$$a < x < b \tag{1}$$

has neither a maximum nor a minimum. However, this set of numbers does have a *least upper bound, b,* and a *greatest lower bound, a.* An *upper bound* of a set of real numbers is a number which is greater than or equal to any number in the set. Similarly a *lower bound* is a number which is less than or equal to any number in the set.

A set of real numbers is said to be *bounded* if there are constants c and k such that the inequalities $c \leq x \leq k$ hold for every number x in the set. The set is said to be *bounded above* if $x \leq k$ holds, and *bounded below* if $c \leq x$ holds, for every x in the set. Any bounded set of real numbers has a unique *least* upper bound and a unique *greatest* lower bound. This statement is not proved here, because for our purposes we need only the very special case where the sets are restricted to be intervals on the real line. (The x axis in analytic geometry is a well-known illustration of the "real line.") For example, the set of numbers x satisfying $a < x < b$ forms an *open interval*, denoted by (a, b).

The set of numbers x satisfying $a \leq x \leq b$ constitutes a *closed interval* denoted by $[a, b]$. This set of numbers has a maximum b, which is also the least upper bound, and a minimum a, which is also the greatest lower bound.

The notation $[a, b)$ denotes the interval consisting of all numbers x satisfying $a \leq x < b$. This set has a minimum a, but no maximum. It has least upper bound b and greatest lower bound a, as also does the set of x satisfying $a < x \leq b$. This latter interval is denoted by $(a, b]$. In this case the set has a maximum, b, but no minimum.

The words "supremum" and "infimum" are often used in the mathematical literature in place of "least upper bound" and "greatest lower bound," but we shall not use these terms.

It is sometimes convenient, in seeking the maximum or the minimum of a function $f(x)$, to look instead for the minimum or maximum of $-f(x)$. For example, if we know that the minimum value of $9 + x^2 - 2x$ over all real numbers x is 8, it follows that the maximum of $2x - 9 - x^2$ is -8. (These results follow readily from the identity $9 + x^2 - 2x = 8 + (x - 1)^2$.) It follows also that the minimum of $90 + x^2 - 2x$ is 89, and the minimum of $900 + 10x^2 - 20x$ is 890.

Reciprocals can be used in the same way. Continuing the example in the preceding paragraph, it follows also that the maximum value of $1/(9 + x^2 - 2x)$ over all real numbers x is $1/8$.

Next we turn to some geometric conventions. For any distinct points P and Q the notation PQ will be used in three senses, easily distinguishable by context: the *straight line* PQ, meaning the infinite line extending in both directions; the *line segment* PQ, namely, the portion of the line terminating at P at one end and Q at the other; and the *distance* PQ, which is a positive number for distinct points P and Q, so that $PQ = QP$. Thus a distance PQ is never negative, and $PQ = 0$ iff the points coincide. (The word "iff" is the shortened form of "if and only if.") As an illustration of the sense being readily determined from the context, such an equation as $PQ = RS$ clearly refers to the equality of two distances.

The *half-line*, or ray, PQ is the line beginning at P as an endpoint and extending from P to Q and indefinitely beyond Q.

A *triangle* consists of three noncollinear points, say A, B, C, together with the line segments AB, BC, AC. Thus the area of a triangle is positive, never zero. The *triangle inequality* states that the sum of the lengths of any two sides exceeds the third, for example, $AB + BC > AC$. More generally, given three distinct points P, Q, R

we have $PQ + QR \geq PR$, with equality iff the point Q lies on the line segment PR. We say that a point Q is an *interior point* of a line segment PR if it lies strictly between P and R on the segment.

For any integer $n \geq 3$, an *n-gon*, or *polygon of n sides*, consists of a set of n distinct points $P_1, P_2, \ldots, P_n$ lying in a plane, called the *vertices*, and the n line segments P_1P_2, P_2P_3, P_3P_4, $\ldots$, $P_{n-1}P_n$, P_nP_1, called the *sides*, satisfying the condition that the sides have no points in common except that each pair of adjacent sides has exactly one vertex in common. The sides collectively form the *perimeter* or *boundary* of the polygon, which effectively separates the exterior points from the interior points.

A polygon is *convex* if the line segment joining any two points on the polygon contains no exterior point, that is, no point lying outside the polygon. Thus a polygon is convex iff each of its interior angles is less than or equal to 180°. More generally, a set S of points is said to be *convex* iff for every pair of points A, B in S the entire line segment AB is contained in S.

1.2. Geometry and Trigonometry. The angle PCQ subtended at the center C of a circle by any arc PQ is twice the angle PKQ subtended by PQ at any point K on the complementary arc, as shown in Figure 1.2a. It follows that $\angle PKQ = \angle PHQ$ for any points H and K lying on the same arc from P to Q. Also, a quadrilateral $PQRS$ is inscribable in a circle if and only if the sum of a pair of opposite angles is 180°. The sum of all four interior angles of a quadrilateral is 360°. The sum of all n interior angles of an n-gon is $180(n - 2)$ degrees.

If P is any point on a semicircle with diameter AB, as shown in Figure 1.2b, then $\angle APB = 90°$. More briefly, the angle in a semicircle is a right angle. Conversely, given any curve from A to B such that $\angle APB = 90°$ for *every* point P on the curve, then *the curve is a semicircle.* This converse is not so widely known, so we give a proof. Impose a coordinate system with AB as the x axis and the origin at the midpoint C of the segment AB. Let c denote the length BC, so that the coordinates of B and A are $(c, 0)$ and $(-c, 0)$. If the coordinates of any point P on the curve are (x, y), then the slopes of the lines PB and PA are

$$\frac{y-0}{x-c} \quad \text{and} \quad \frac{y-0}{x+c}.$$

These are perpendicular lines, so the product of the slopes is -1, giving

$$\frac{y}{x-c}\cdot\frac{y}{x+c} = -1, \qquad \frac{y^2}{x^2-c^2} = -1.$$

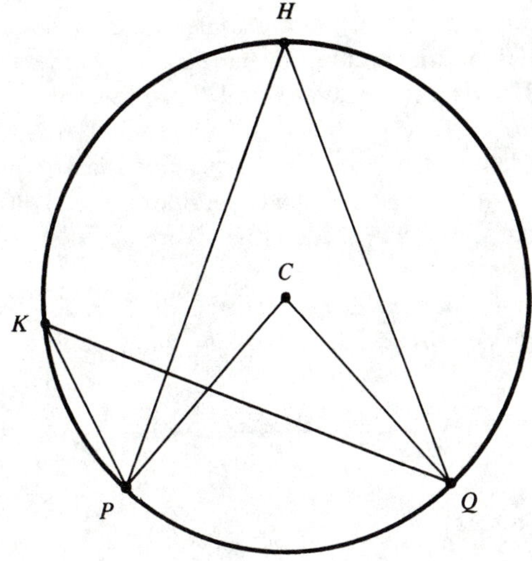

FIG. 1.2a

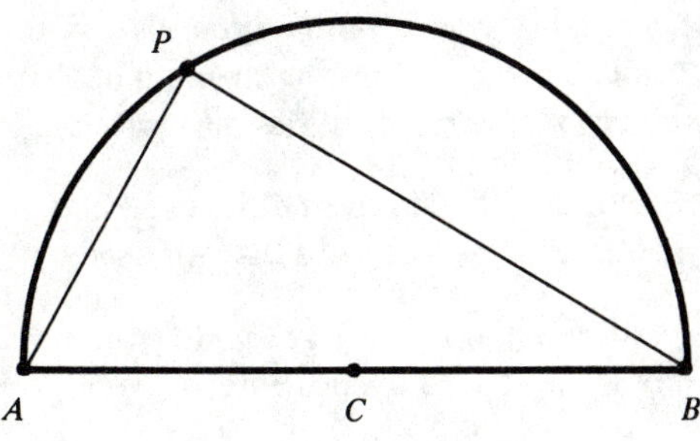

FIG. 1.2b

Multiplying by $x^2 - c^2$ we get $y^2 = -x^2 + c^2$ or $x^2 + y^2 = c^2$. This is the well-known equation of a circle of radius c, with center at the origin. Each of the two semicircles based on the segment AB has the property $\angle APB = 90°$ for every point P on the curve.

The triangle inequality states that if a, b, c are the lengths of the sides of any triangle then

$$a + b > c, \quad a + c > b, \quad b + c > a.$$

Also if two sides of a triangle are unequal, then the angles opposite these sides are unequal in the same sense; that is, if $a > b$ then $\alpha > \beta$, and conversely, where α and β are the angles opposite the sides a and b.

The triangle inequality has an obvious extension to quadrilaterals, that the sum of the lengths of any three sides exceeds the fourth, and to n-gons, that the sum of the lengths of any $n - 1$ sides exceeds the length of the nth side.

Let the lengths of the sides of any triangle be a, b, c, and let α, β, γ be the angles opposite. The *law of sines* states that

$$\frac{\sin \alpha}{a} = \frac{\sin \beta}{b} = \frac{\sin \gamma}{c}.$$

The *law of cosines* is

$$c^2 = a^2 + b^2 - 2ab \cos \gamma,$$

along with two analogous equations with c, a, b, γ replaced by a, b, c, α and by b, c, a, β. In case $\gamma = 90°$, then $\cos \gamma = 0$ and the equation above becomes $c^2 = a^2 + b^2$, the theorem of Pythagoras.

The following trigonometric identities are useful:

$$\sin(\alpha \pm \beta) = \sin \alpha \cos \beta \pm \cos \alpha \sin \beta,$$
$$\cos(\alpha \pm \beta) = \cos \alpha \cos \beta \mp \sin \alpha \sin \beta,$$

where in each equation the upper signs go together and likewise the lower signs. Simple corollaries of these are:

$$\sin 2\theta = 2 \sin \theta \cos \theta,$$
$$\cos 2\theta = \cos^2 \theta - \sin^2 \theta,$$

and

$$\sin\alpha + \sin\beta = 2\sin\frac{\alpha+\beta}{2}\cos\frac{\alpha-\beta}{2}.$$

1.3. Areas and Volumes. There are three standard formulas for the area of a triangle, each useful in its own way. If a triangle has sides of lengths a, b, c and angles α, β, γ opposite these sides, then the area A is given by

$$A = \frac{1}{2}ab\sin\gamma = \frac{1}{2}ac\sin\beta = \frac{1}{2}bc\sin\alpha. \tag{1}$$

This is just one formula, in three variations. A second basic result is $A = bh/2$ where h is the length of the altitude or height of the triangle drawn to the side b. Not so well known perhaps is *Heron's, or Hero's, formula*

$$A = \{s(s-a)(s-b)(s-c)\}^{1/2} \tag{2}$$

where s is the *semi-perimeter*; that is, $s = (a+b+c)/2$. The proof of this is quite easy, as follows. First we square the earlier formula (1) to get

$$A^2 = \frac{1}{4}a^2b^2\sin^2\gamma = \frac{1}{4}a^2b^2(1-\cos^2\gamma).$$

Replace $\cos^2\gamma$ by its value from the law of cosines to obtain

$$\begin{aligned} A^2 &= \frac{1}{4}a^2b^2\left[1 - \frac{(a^2+b^2-c^2)^2}{4a^2b^2}\right] \\ &= \frac{1}{16}[4a^2b^2 - (a^2+b^2-c^2)^2]. \end{aligned}$$

Inside the square brackets we have the difference of two squares, giving the usual factoring:

$$16A^2 = [2ab + (a^2+b^2-c^2)]\cdot[2ab - (a^2+b^2-c^2)].$$

Removing the parentheses, we find that inside each pair of square brackets again there is a difference of two squares, giving the further factoring:

$$\begin{aligned}16A^2 &= (a+b+c)(a+b-c)(a-b+c)(-a+b+c)\\ &= 2s(2s-2c)(2s-2b)(2s-2a).\end{aligned}$$

Dividing by 16 and taking the square root, we get formula (2).

There is a similar formula for the area A of a quadrilateral,

$$A^2 = (s-a)(s-b)(s-c)(s-d) - \frac{1}{2}abcd\,[1+\cos(\theta+\lambda)], \quad (3)$$

where a, b, c, d are the lengths of the sides, s is the semiperimeter so that $s = (a+b+c+d)/2$, and θ and λ are two opposite interior angles of the quadrilateral. It does not matter which pair of opposite angles are labeled θ and λ in this formula, because if the other pair is α, β we have

$$(\theta+\lambda)+(\alpha+\beta) = 360^\circ \quad \text{and} \quad \cos(\theta+\lambda) = \cos(\alpha+\beta)$$

by elementary trigonometry.

If the quadrilateral is inscribable in a circle we have $\theta + \lambda = 180^\circ$, and hence $\cos(\theta+\lambda) = -1$ and the formula for the area A can be written

$$A = \{(s-a)(s-b)(s-c)(s-d)\}^{1/2}. \quad (4)$$

To prove formula (3) we draw a diagonal, say of length x, as in Figure 1.3a with the angle θ between sides a and b, and λ between c and d. The law of cosines gives

$$x^2 = a^2 + b^2 - 2ab\cos\theta = c^2 + d^2 - 2cd\cos\lambda, \quad (5)$$

and so

$$a^2 + b^2 - c^2 - d^2 = 2ab\cos\theta - 2cd\cos\lambda.$$

Squaring both sides here and regrouping we get

$$(a^2+b^2-c^2-d^2)^2 + 8\,abcd\cos\theta\cos\lambda$$
$$= 4a^2b^2\cos^2\theta + 4c^2d^2\cos^2\lambda. \quad (6)$$

The area A of the quadrilateral is seen to be, from (1),

$$A = \frac{1}{2}ab\sin\theta + \frac{1}{2}cd\sin\lambda.$$

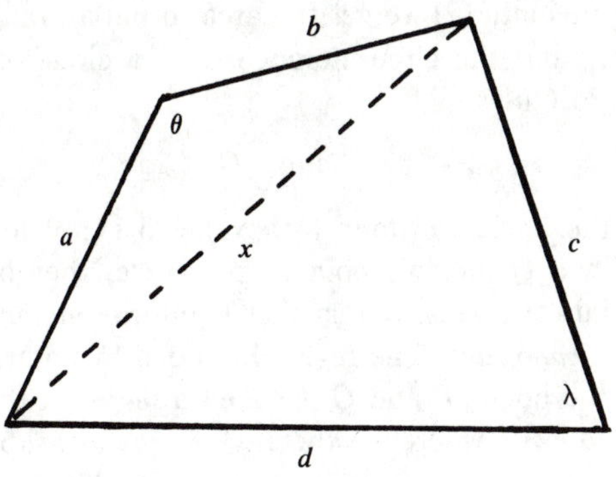

FIG. 1.3a

Squaring this and multiplying by 16 we get

$$\begin{aligned}16A^2 &= 4a^2b^2\sin^2\theta + 4c^2d^2\sin^2\lambda + 8abcd\sin\theta\sin\lambda\\ &= 4a^2b^2 + 4c^2d^2 - 4a^2b^2\cos^2\theta - 4c^2d^2\cos^2\lambda\\ &\qquad + 8abcd\sin\theta\sin\lambda.\end{aligned}$$

Using (6) above we see that

$$\begin{aligned}16A^2 &= 4a^2b^2 + 4c^2d^2 - (a^2+b^2-c^2-d^2)^2\\ &\qquad - 8abcd\cos\theta\cos\lambda + 8abcd\sin\theta\sin\lambda\\ &= 4a^2b^2 + 4c^2d^2 - (a^2+b^2-c^2-d^2)^2\\ &\qquad - 8abcd\cos(\theta+\lambda).\end{aligned}$$

Next we add and subtract the term $8abcd$ to get

$$\begin{aligned}16A^2 = (2ab+2cd)^2 &- (a^2+b^2-c^2-d^2)^2\\ &- 8abcd[1+\cos(\theta+\lambda)]. \qquad (7)\end{aligned}$$

The first two terms on the right side form the difference of two squares which can be factored, and each factor again turns out to be the difference of two squares. Thus we can write

$$\begin{aligned}&(2ab+2cd)^2 - (a^2+b^2-c^2-d^2)^2\\ &= (2ab+2cd+a^2+b^2-c^2-d^2)\\ &\qquad \times (2ab+2cd-a^2-b^2+c^2+d^2)\\ &= (a+b+c-d)(a+b-c+d)\\ &\qquad \times (a-b+c+d)(-a+b+c+d)\\ &= (2s-2d)(2s-2c)(2s-2b)(2s-2a).\end{aligned}$$

Substituting this into (7) we get the area formula (3).

The area A and the circumference C of a circle of radius r are given by the formulas

$$A = \pi r^2 \quad \text{and} \quad C = 2\pi r,$$

where π is the basic constant with value 3.14159 to five decimal places. If P and Q are two points on a circle, they divide the circumference into two arcs, the smaller being the *minor arc PQ*, and the larger the *major arc*. The region bounded by an arc PQ and the two radii to the points P and Q is called a *sector* of the circle. This sector has area $\frac{1}{2}r^2\theta$, where θ is the angle in *radians* subtended at the center by the arc PQ, as shown in Figure 1.3b. The length of the arc PQ is $r\theta$. The basic relation between degree measure and radian measure of angles is $180° = \pi$ radians.

The volume V and the surface area S of a sphere of radius r are given by

$$V = \frac{4}{3}\pi r^3 \quad \text{and} \quad S = 4\pi r^2.$$

For a right circular cylinder (the shape of a tin can) the formulas are

$$V = \pi r^2 h \quad \text{and} \quad S = 2\pi r^2 + 2\pi r h,$$

where r is the radius of the circular base and h is the height.

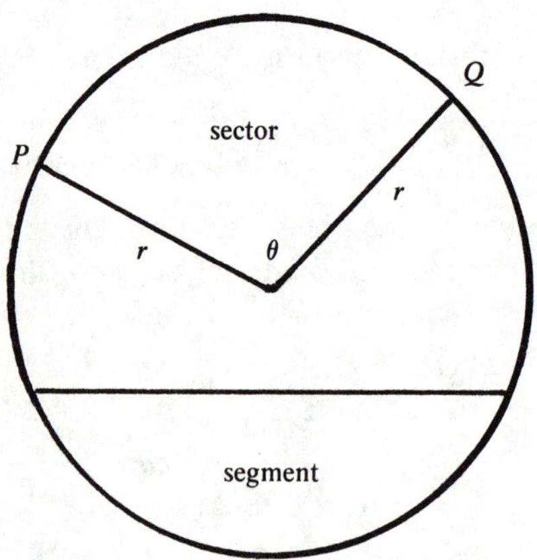

FIG. 1.3b

For a right circular cone the volume and surface area are

$$V = \frac{1}{3}\pi r^2 h \quad \text{and} \quad S = \pi r^2 + \pi r(r^2 + h^2)^{1/2}$$

where πr^2 is of course the area of the circular base and the other term in S is the lateral area.

The volume of a tetrahedron is $\frac{1}{3}Ah$, where A is the area of any triangular face and h is the length of the perpendicular drawn from the fourth vertex to that face.

A1. If a quadrilateral inscribed in a circle has sides of lengths a, b, c, d, and if $a + b = c + d$, prove that its area is $(abcd)^{1/2}$.

A2. Is formula (3) valid for the area of a nonconvex quadrilateral, that is, a quadrilateral one of whose diagonals lies outside the figure?

A3. Given any triangle, prove that it is possible to define the unit of length so that the area of the triangle is 1. (The purpose of this problem is to enable us to write, in certain problems, "Without loss of generality we can take the area of the triangle to be 1, by redefining the unit of length if necessary.")

1.4. Inequalities. To supplement the basic definitions of inequalities given in §1.1, we now state some fundamental results without proof.

The *transitive* property: if $a > b$ and $b > c$, then $a > c$. This result is valid if the assumption $a > b$ is replaced by the weaker version $a \geq b$.

If a, b, c, d, and k are positive real numbers satisfying $a \geq b$ and $c > d$, then

(1) $a + c > b + d$ and $ac > bd$,
(2) $a + k \geq b + k$ and $ak \geq bk$,
(3) $c + k > d + k$ and $ck > dk$,
(4) $a^{-1} \leq b^{-1}$ and $c^{-1} < d^{-1}$,
(5) $a^n \geq b^n$ and $c^n > d^n$,
(6) $a^{1/n} \geq b^{1/n}$ and $c^{1/n} > d^{1/n}$,

where n is any positive integer.

The first inequality in each of (1), (2), (3) holds without the restriction that the numbers are positive.

The *absolute value* of a real number r, denoted by $|r|$, may be defined as $\max(r, -r)$. An alternative definition is that $|r| = r$ in case $r \geq 0$, whereas $|r| = -r$ in case $r < 0$.

In the following problems a, b, c, d denote positive real numbers.

A4. Does $a + b > c$ imply $a^2 + b^2 > c^2$, or $\sqrt{a} + \sqrt{b} > \sqrt{c}$?

A5. Given the information $a + b > c + d$, $a > b$ and $c > d$, does this imply any of $a > c$, $a > d$, $b > c$, or $b > d$? Give a proof or a counterexample in each case.

A6. Find $\max(a, a^2)$ and $\max(\sqrt{a} + \sqrt{b}, \sqrt{a+b})$.

A7. Which of the following, if any, hold for all real numbers r, s, and t?

(i) $[\max(r, s, t)]^2 = \max(r^2, s^2, t^2)$,
(ii) $[\max(r, s, t)]^3 = \max(r^3, s^3, t^3)$,
(iii) $|\max(r, s, t)| = \max(|r|, |s|, |t|)$.

A8. If α, β, and γ denote the angles of any triangle, can we conclude that $\sin\alpha + \sin\beta > \sin\gamma$?

A9. *Newton's approximation process for square roots.* For any positive number c, let x_1 be an approximation to $\sqrt{c}$, say with $x_1 \neq \sqrt{c}$. Define $x_2, x_3, x_4, \ldots$ successively by

$$x_2 = \frac{1}{2}(x_1 + c/x_1), \quad x_3 = \frac{1}{2}(x_2 + c/x_2), \ldots,$$

$$x_{n+1} = \frac{1}{2}(x_n + c/x_n), \ldots.$$

For example if $c = 2$ and x_1 is taken as 1, then $x_2 = 3/2$, $x_3 = 17/12$ and $x_4 = 577/408 = 1.4142157\ldots$, which is accurate to five decimal places as an approximation to $\sqrt{2}$. Prove that $x_n > \sqrt{c}$ if $n > 1$. Prove that if $n > 1$, then x_{n+1} is closer to $\sqrt{c}$ than x_n is, in fact, that $x_{n+1} - \sqrt{c} < \frac{1}{2}(x_n - \sqrt{c})$.

A10. For any triples of real numbers $A = \{a_1, a_2, a_3\}$ and $B = \{b_1, b_2, b_3\}$ say that $A > B$ if at least five of the nine inequalities

$a_i > b_j$ hold, where $i = 1, 2, 3$ and $j = 1, 2, 3$. With inequality between triples defined in this way, does the transitive property hold, that is, does $A > C$ follow from $A > B$ and $B > C$? If so, prove it; if not, produce a counterexample.

1.5. The Sigma Notation. We shall occasionally use sums like $a_1 + a_2 + a_3 + \cdots + a_9$ or $x_1y_1 + x_2y_2 + x_3y_3 + \cdots + x_ny_n$. It is customary to denote the individual terms of such sums by using a general subscript, such as a_i with $i = 1, 2, 3, \ldots, 9$, or x_jy_j with $j = 1, 2, 3, \ldots, n$. The sums themselves can then be written in the sigma notation

$$\sum_{i=1}^{9} a_i = a_1 + a_2 + a_3 + \cdots + a_9,$$

and

$$\sum_{j=1}^{n} x_jy_j = x_1y_1 + x_2y_2 + x_3y_3 + \cdots + x_ny_n.$$

Simple extensions follow at once from elementary algebra, such as

$$\sum_{i=1}^{9} ca_i = c\sum_{i=1}^{9} a_i, \quad \sum_{i=1}^{9} (c + a_i) = 9c + \sum_{i=1}^{9} a_i,$$

and

$$\sum_{1}^{n} (x_j - y_j)^2 = \sum_{1}^{n} x_j^2 - 2\sum_{1}^{n} x_jy_j + \sum_{1}^{n} y_j^2.$$

In the last equation the index j has been omitted because the meaning is clear as it stands. Similarly, $\Sigma_1^{100} j^2$ is a brief way of writing the sum of the squares of the natural numbers from 1 to 100, namely, the sum $1^2 + 2^2 + 3^2 + \cdots + 100^2$.

Notes on Chapter 1

§1.2. Excellent discussions of the fundamental concepts of geometry are given in the following books: Coxeter and Greitzer (1967); Eves (1972); Moise (1963); Pedoe (1970). See the list of references on page 293 for details.

If a quadrilateral is inscribed in a circle, then $ac + bd = pq$, where a, b, c, d are the lengths of the sides in that order, and p, q the lengths of the diagonals. This is Ptolemy's theorem, an extended version of which is given in §10.5.

§1.4. See the book by Beckenbach and Bellman (1961) for a thorough discussion of elementary inequalities, including proofs of the basic results.

CHAPTER **2**

SIMPLE ALGEBRAIC RESULTS

2.1. Sums and Products. Consider the question of finding two numbers whose sum is 60 and whose product is as large as possible. The answer is 30 and 30. Any other pair with sum 60, such as 20 and 40, has a smaller product. A natural extension is to ask for three positive numbers whose sum is 60 and having the largest possible product. The answer is now 20, 20, and 20. For four positive numbers with sum 60, the largest product occurs with the numbers 15, 15, 15, and 15. The idea seems clear: make the numbers equal. This is one of the central themes of this chapter, and the analogous idea with sums and products reversed is another basic concept.

Sums and products reversed: To be more specific, the question now is to find two positive numbers whose product is, say, 64, and whose sum is a *minimum.* The answer is 8 and 8. Why do we ask for a minimum sum? Why not a maximum? A moment's reflection reveals that no maximum sum can be attained, because the sum can be made as large as we please. For example, the numbers 64000 and .001 have product 64 and a modestly large sum. Pairs of numbers with larger sums, but still with product 64, can easily be created in the obvious way.

The natural extension is to ask for three positive numbers with product 64 and with a minimum sum. Again we seek three equal numbers, so the answer is 4, 4, and 4.

The general ideas described above are developed with proofs in this chapter, and applied to various algebraic problems. Geometric applications are considered in the next chapter. The reasoning underlying these results is based on the inequality of the arithmetic-geometric means. The *arithmetic mean,* or arithmetic average, of a set of numbers is the best known of all averages: if there are n numbers, add them and divide by n. The arithmetic mean of $a_1, a_2, \ldots, a_n$ is thus $(a_1 + a_2 + \cdots + a_n)/n$, denoted by A. The *geometric mean,* usually restricted to nonnegative numbers, is $(a_1 a_2 \cdots a_n)^{1/n}$, denoted by G. The AM-GM inequality, or inequality of the arithmetic-geometric means, states that $A \geq G$, with equality iff the numbers $a_1, a_2, \ldots, a_n$ are all equal. This important inequality is developed in the first few sections, with its applications explored in §2.6.

Other kinds of averages are introduced in §2.7, along with the related inequalities. One nice byproduct of the theory of this chapter is the analysis of the basic mathematical constant e in §2.8.

2.2. Any Square Is Positive or Zero. The square of any real number is positive or zero. Furthermore, the square is zero iff the number itself is zero. From these simple concepts some basic results are quickly obtained.

THEOREM 2.2a. *For any constant* c, *the maximum value of* $cx - x^2$ *over all real numbers* x *is* $c^2/4$, *occurring iff* $x = c/2$.

This can be seen by writing

$$cx - x^2 = \frac{c^2}{4} - \left(x - \frac{c}{2}\right)^2, \tag{1}$$

an identity easily verified. Since $(x - c/2)^2$ is positive or zero, the largest value of $cx - x^2$ is $c^2/4$, obtained only by choosing $x = c/2$.

There is no *minimum* value of $cx - x^2$, as is easily seen by considering what happens if x gets very large. The theorem could have been stated with a more general coefficient for x^2, but such a case is easily reduced to $cx - x^2$. For example, to find the largest value of $24x - 4x^2$, we write this as $4(6x - x^2)$ and apply the theorem to observe that the maximum occurs at $x = 3$. Thus the maximum is 36. Similarly, an added constant is easily dealt with. To find the maximum value of $50 + 24x - 4x^2$, we ignore the constant 50 for a moment and look for the largest value of $24x - 4x^2$. As noted above, this largest value is 36, so the maximum of $50 + 24x - 4x^2$ is 86.

COROLLARY 1. *The minimum value of* $x^2 - cx$ *is* $-c^2/4$, *occurring iff* $x = c/2$.

This follows at once because $x^2 - cx$ is the negative of $cx - x^2$, so that the largest value of $cx - x^2$ and the smallest value of $x^2 - cx$ occur iff $x = c/2$.

COROLLARY 2. *If two variables* x *and* y *satisfy* $x + y = c$, *their product* xy *is a maximum iff the numbers are equal, that is,* $x = y = c/2$.

Using $y = c - x$ we note that $xy = cx - x^2$, so this result follows at once from the theorem.

COROLLARY 3. *If two positive variables satisfy* $xy = c$, *where* c *is a positive constant, their sum* $x + y$ *is a minimum iff* $x = y = \sqrt{c}$.

This is a corollary in the sense that, as in the proof of the theorem, the basic idea is to *complete a square.* Writing $y = c/x$ we see that

$$x + y = x + c/x = (\sqrt{x} - \sqrt{c/x})^2 + 2\sqrt{c}.$$

The minimum value of $x + y$ occurs if the square is zero, that is, $\sqrt{x} = \sqrt{c/x}$ which gives $x = \sqrt{c}$ if we multiply by $\sqrt{x}$. Then $y = \sqrt{c}$ follows.

Note that it is specified in this result that the variables x and y must be positive. This is important, because without this restriction the sum $x + y$ has no minimum. For example if $xy = 25$, the corollary assures us that the least value of $x + y$ is 10 provided x and y are restricted to positive values. If negative numbers are allowed, we can make $x + y$ as small as we please.

Example 1. Consider the sum of any positive real number and its reciprocal. Prove that the least value of this sum is 2.

Solution. Writing x for the positive real number, we want to minimize $x + 1/x$. This is the sum of two terms whose product is a constant; so Corollary 3 applies. The sum is a minimum when x and $1/x$ are equal, from which we get

$$x = 1/x, \quad x^2 = 1, \quad x = 1,$$

where we have discarded the solution $x = -1$ because x is positive. Thus the minimum of $x + 1/x$ is 2.

Example 2. Let a and b be positive constants. Find the least value of $ax + b/x$ for all positive numbers x.

Solution. Again we have a sum of two terms whose product is constant; so we can apply Corollary 3. The sum is minimized if the terms are equal; that is,

$$ax = b/x, \quad x^2 = b/a, \quad x = \sqrt{b/a}.$$

The minimum value of the sum is therefore $2\sqrt{ab}$.

Example 3. Let a, b, and c be positive constants. Among all real numbers x and y satisfying $ax + by = c$, find the maximum value of the product xy.

Solution. If we write $u = ax$, $v = by$, then $ax + by = c$ becomes $u + v = c$. By Corollary 2 the product uv is maximized by taking $u = v = c/2$. But we see that

$$uv = abxy \quad \text{or} \quad xy = uv/ab.$$

Hence to maximize uv is to maximize xy also, and this occurs in case

$$u = v = c/2 = ax = by, \quad x = c/2a, \quad y = c/2b.$$

Thus the maximum value of xy is $c^2/4ab$.

THEOREM 2.2b. *For any real numbers a and b the inequalities*

$$a^2 + b^2 \geq 2ab, \quad \left(\frac{a+b}{2}\right)^2 \geq ab, \quad (a+b)^2 \geq 4ab$$

hold, with equality in each case iff $a = b$. If a and b are nonnegative real numbers, then

$$\frac{a+b}{2} \geq \sqrt{ab},$$

with equality iff $a = b$.

The last result here is the inequality of the arithmetic-geometric means for two positive numbers. It is clear that we need the restriction that a and b are not negative, because in the case $a = b = -6$, for example, the inequality $(a + b)/2 \geq \sqrt{ab}$ is incorrect.

The proof of this theorem is straightforward because the first three inequalities are simply variations on

$$(a-b)^2 \geq 0, \qquad a^2 - 2ab + b^2 \geq 0.$$

Furthermore we know that the strict inequality $(a-b)^2 > 0$ holds unless $a = b$.

Similarly, the inequality $(a+b)/2 \geq \sqrt{ab}$ can be derived from

$$(\sqrt{a} - \sqrt{b})^2 \geq 0 \quad \text{or} \quad a - 2\sqrt{ab} + b \geq 0.$$

Let us return for a moment to Theorem 2.2a, wherein it is observed that the maximum of $cx - x^2$ is $c^2/4$. Now the method of proof of this focused on $cx - x^2$ as a quadratic polynomial in x. But of far greater significance for purposes of wider application is the argument that views $cx - x^2$ as a product of two factors x and $c - x$ having the constant sum c. Then the maximum occurs if the two factors are equal. Suppose we have a product of three positive factors having the property that the sum of the factors is a positive constant. Is the product a maximum when the factors are equal? The answer is yes, and moreover the same idea applies for four factors, or five, or any finite number. These extensions of Theorem 2.2a are given in the next few sections.

B1. Let a, b, and c be positive constants. For all positive numbers x and y with product c, find the minimum value of $ax + by$.

B2. If a, b, c are real numbers, not all equal, prove that

$$a^2 + b^2 + c^2 > ab + ac + bc.$$

B3. Find the real number that exceeds its square by the largest amount.

B4. The *harmonic mean* of two positive numbers a and b is the reciprocal of the arithmetic mean of their reciprocals, that is,

$$\left[\left(\frac{1}{a} + \frac{1}{b}\right)\Big/2\right]^{-1}.$$

Prove that the harmonic mean can be written as $2ab/(a+b)$. If $a \neq b$, prove that the harmonic mean is less than the geometric mean, so that we have

$$\frac{a+b}{2} > \sqrt{ab} > \frac{2ab}{a+b}.$$

If $a = b$, these three means are seen to be equal.

B5. Given any positive constant c, find the minimum value of $x^4 + 2y^4$ for positive numbers x and y having product $xy = c$.

B6. For any constants a, b, c find the minimum value of

$$x^2 + y^2 + ax + by + c$$

over all real numbers x and y.

B7. For any positive constant c, prove that $2cx - x^2$ decreases but $x + c^2/x$ increases as x increases from c through larger positive values. Prove analogous results in case x decreases from c through smaller positive values.

B8. If x, y, r, s are any real numbers satisfying $x^2 + y^2 = 1$ and $r^2 + s^2 = 1$, find the maximum of $xr + ys$.

B9. Find the largest possible value of xy if x and y are subject to the condition $20x + y = 180$.

B10. A heavy trailer-truck makes a regular highway trip of 400 miles at almost constant speed. The vehicle laws require that the speed be not less than 35 and not more than 55 miles per hour. The truck burns fuel at a rate of $1 + x/40 + x^2/300$ gallons per hours at a speed of x miles per hour. If the cost of the fuel is k dollars per gallon, and the driver's wages are $8k$ dollars per hour, at what speed should the truck be driven to minimize the total cost of the trip, including fuel costs and driver's wages?

2.3 The Inequality of the Arithmetic-Geometric Means. In this section we prove a fundamental inequality having extensive applications to extremal problems. The inequality is $A \geq G$, where A is the arithmetic mean, and G the geometric mean, of a finite collection of nonnegative numbers. For the special case of two numbers, say a and b, the result takes the form $(a + b)/2 \geq \sqrt{ab}$, proved in Theorem 2.2b. The proof given below of the general result $A \geq G$ is conceptually very simple, not as short as some other available proofs, but very easy to follow. In the next two sections we give a different proof, using an argument that can be adapted to more general situations in mathematics, as we shall see in Chapter 5.

THEOREM 2.3a. *The Inequality of the Arithmetic-Geometric Means. Let A and G denote the arithmetic and geometric means of the nonnegative numbers $a_1, a_2, \ldots, a_n$ defined by*

$$A = \frac{a_1 + a_2 + \cdots + a_n}{n}, \qquad G = (a_1 a_2 \cdots a_n)^{1/n}. \tag{1}$$

Then $A \geq G$, with equality iff $a_1 = a_2 = \cdots = a_n$.

In case $a_1 = a_2 = \cdots = a_n$ it is clear from (1) that $A = G$. Henceforth we assume that the a's are not all equal, and prove that $A > G$. Actually we prove the equivalent version $G^n < A^n$; that is,

$$a_1 a_2 \cdots a_n < A^n. \tag{2}$$

This inequality is established by replacing the product on the left by successively larger products, reaching A^n in fewer than n steps. Each step in the process is described by the following *algorithm,* or procedure.

ALGORITHM. In any product of n numbers, replace the smallest number, say x, and the largest number, say y, by two new factors A and $x + y - A$, where A denotes the arithmetic mean of the n numbers.

We illustrate this with two examples.

Example 1. The inequality $2 \cdot 3 \cdot 4 \cdot 6 \cdot 20 < 7^5$, a special case of (2) above, can be obtained by repeated application of the algorithm:

$$\begin{aligned} 2 \cdot 3 \cdot 4 \cdot 6 \cdot 20 &< 3 \cdot 4 \cdot 6 \cdot 7 \cdot 15 \\ &< 4 \cdot 6 \cdot 7 \cdot 7 \cdot 11 \\ &< 6 \cdot 7 \cdot 7 \cdot 7 \cdot 8 \\ &< 7 \cdot 7 \cdot 7 \cdot 7 \cdot 7. \end{aligned}$$

The arithmetic mean of the five numbers in each product is 7. In the first step the smallest and largest numbers, 2 and 20, are replaced by 7 and 15, because $x + y - A = 2 + 20 - 7 = 15$. In the second step the smallest and largest numbers, 3 and 15, are replaced by 7 and 11, because now we have $x + y - A = 3 + 15 - 7 = 11$. The subsequent steps again follow the algorithm. (Of course, the inequality $2 \cdot 3 \cdot 4 \cdot 6 \cdot 20 < 7^5$ can be verified by a direct calculation. The general inequality (2) *cannot* be verified in such a simple way, but it can be obtained by use of the algorithm, as we shall see.)

Example 2. The inequality $1 \cdot 6 \cdot 7 \cdot 10 \cdot 11 \cdot 19 < 9^6$, another special case of the inequality (2) above, can be obtained by repeated application of the algorithm:

$$\begin{aligned} 1 \cdot 6 \cdot 7 \cdot 10 \cdot 11 \cdot 19 &< 6 \cdot 7 \cdot 9 \cdot 10 \cdot 11 \cdot 11 \\ &< 7 \cdot 8 \cdot 9 \cdot \ 9 \cdot 10 \cdot 11 \\ &< 8 \cdot 9 \cdot 9 \cdot \ 9 \cdot \ 9 \cdot 10 \\ &< 9 \cdot 9 \cdot 9 \cdot \ 9 \cdot \ 9 \cdot \ 9. \end{aligned}$$

We note that the algorithm replaces the numbers x and y, the smallest and the largest in any product, by two numbers A and $x + y - A$ having the same sum. Hence the algorithm, when applied to $a_1 \, a_2 \, \cdots \, a_n$, replaces a product of n factors having arithmetic mean A by another product of n factors also having arithmetic mean A. Why is the new product larger? To answer this question, we prove that

$$xy < A(x + y - A). \tag{3}$$

Removing the parentheses, and moving the term xy to the right side of the inequality, we see that (3) is equivalent to

$$0 < Ax + Ay - A^2 - xy \quad \text{or} \quad 0 < (A - x)(y - A). \tag{4}$$

This final inequality is easily verified because $A - x$ and $y - A$ are both positive. The reason for this is that the arithmetic average of n numbers lies between the smallest and the largest of the numbers, that is, $x < A < y$. This verifies (4).

Finally we see that repeated application of the algorithm to the product $a_1 \, a_2 \, \cdots \, a_n$ leads to the product A^n, because each step brings in one or two more occurrences of the factor A, as illustrated in the examples above. This proves the inequality (2), and so also Theorem 2.3a.

COROLLARY 1. *If $a_1, a_2, \ldots, a_n$ are nonnegative, then $a_1{}^n + a_2{}^n + \cdots + a_n{}^n \geq n a_1 \, a_2 \cdots a_n$, with equality iff the a's are all equal.*

This follows by applying Theorem 2.3a to the numbers $a_1{}^n$, $a_2{}^n$, $\ldots$, $a_n{}^n$, with $G = a_1 \, a_2 \cdots a_n$ in this case.

B11. Construct a dual version of the proof given above of the AM-GM inequality, along the following lines. Setting aside the case where $a_1 = a_2 = \cdots = a_n$, prove $A > G$ in the form $nA > nG$, or

$$a_1 + a_2 + \cdots + a_n > G + G + \cdots + G,$$

replacing the sum on the left by $G + G + \cdots + G$ after a succession of steps. At each step replace the smallest number x in the sum and the largest number y by G and xy/G. Note that this replacement leaves the geometric mean G unaltered. For example, the numbers in the sum $3 + 6 + 8 + 144$ have geometric means $G = 12$, so the numbers 3 and 144 are replaced by 12 and 36, giving the sum $6 + 8 + 12 + 36$. Verify in general that this process replaces any sum by a smaller sum.

2.4. An Alternative Approach. In the next section we give what is perhaps the best-known proof, that by Cauchy, of the inequality of the arithmetic-geometric means. In this section we give the special cases of Cauchy's proof for $n = 3$ and 4. Since these special cases are included in the work of the next section, why do we treat them here? The answer is that the Cauchy argument is perhaps a little formidable for a beginning reader, so the easy cases are introduced here as simple background. First we prove

$$(a + b + c + d)/4 \geq (abcd)^{1/4}, \tag{1}$$

with equality iff the nonnegative numbers a, b, c, d are all equal. It is clear that (1) is an equality in case $a = b = c = d$, so we presume that they are not all equal, say $a \neq b$, and prove a strict inequality. From Theorem 2.2b we have

$$a + b > 2\sqrt{ab} \quad \text{and} \quad c + d \geq 2\sqrt{cd}. \tag{2}$$

Adding these and dividing by 2 we get

$$(a + b + c + d)/2 > \sqrt{ab} + \sqrt{cd}. \tag{3}$$

Using Theorem 2.2b again we have

$$\sqrt{ab} + \sqrt{cd} \geq 2\sqrt[4]{abcd}. \tag{4}$$

Note that (1) follows from (3) and (4). Moreover, since (3) is a strict inequality, we get (1) in strict form.

Next we want to prove

$$(a + b + c)/3 \geq (abc)^{1/3}, \tag{5}$$

with equality iff $a = b = c$. Again we assume $a \neq b$ and derive a strict inequality. Replace d by $(abc)^{1/3}$ in (1) to get

$$[a + b + c + (abc)^{1/3}]/4 > [abc(abc)^{1/3}]^{1/4}.$$

The right side reduces to $(abc)^{1/3}$, and multiplication by 4 gives

$$a + b + c + (abc)^{1/3} > 4(abc)^{1/3}.$$

This amounts to (5) with strict inequality.

The arguments in this and the next section originated with the French mathematician Augustin Cauchy (1789–1857) in his *Cours d'analyse,* published in 1821. This book outlines the foundations of calculus in a form that is very little changed today. Cauchy, one of the pioneers in the insistence on rigor in mathematical arguments, was preeminent in several branches of the subject, notably infinite series, differential equations, and complex function theory. A hard worker, he was so productive that the editors of the French mathematical journal *Comptes Rendus* imposed a size limitation on papers for publication, to cope with Cauchy's prolific output.

2.5. Cauchy's Proof. We generalize the argument of the preceding section to give another proof of the AM-GM inequality. The proof is by mathematical induction, actually by induction both upwards and downwards. Let P_n denote the proposition that if a_1, $a_2, \ldots, a_n$ are nonnegative numbers, not all equal, then

$$a_1 + a_2 + \cdots + a_n > n(a_1 a_2 \cdots a_n)^{1/n}. \tag{1}$$

The starting point for a proof by induction is that P_2 holds, by Theorem 2.2b. We view the proofs of P_3 and P_4 in the preceding section as a practice run, and so we establish P_n for all $n \geq 3$. This we do by proving two results.

First Result. *If the proposition P_n holds for any integer $n \geq 3$, then P_{n-1} also holds.*

Second Result. *If the proposition P_n holds for any integer $n \geq 2$, then P_{2n} also holds.*

Before proving these two results, we show how they establish P_n for all n. We know that P_2 holds, so by repeated application of the second result we see that all of

$$P_4, P_8, P_{16}, P_{32}, P_{64}, P_{128}, \ldots \tag{2}$$

hold. All we have to do now is fill in the interstices between any consecutive pair of this sequence. This is achieved by using the first result. For example, suppose we want to establish the truth of P_{57}. Applying the first result with $n = 64$, we see that P_{64} implies P_{63}, and then using $n = 63$, we conclude that P_{63} implies P_{62}. Continuing, we apply the first result with $n = 62, 61, 60, 59, 58$, where the last step here gives us the information that P_{58} implies P_{57}. Thus the two results above are used, in effect by a two-way induction process—first induction up as in (2), and then induction down as explained—to prove P_n for all $n \geq 3$.

Proof that P_n implies P_{n-1}. A model for what we are about to do can be found in the preceding section, where it is proved that P_4 implies P_3. Without loss of generality we may presume that $a_1 \neq a_2$. We are given (1), in which we replace a_n by g, the geometric mean of $a_1, a_2, \ldots, a_{n-1}$, that is,

$$g = (a_1 a_2 \cdots a_{n-1})^{1/(n-1)}. \tag{3}$$

The outcome is

$$a_1 + a_2 + \cdots + a_{n-1} + g > n(a_1 a_2 \cdots a_{n-1} g)^{1/n}. \tag{4}$$

The right side here is

$$n(g^{n-1} g)^{1/n} = n(g^n)^{1/n} = ng.$$

Hence (4) can be written as

$$a_1 + a_2 + \cdots + a_{n-1} + g > ng.$$

Subtracting g from both sides we get (1) with n replaced by $n - 1$, namely, the proposition P_{n-1}.

Proof that P_n implies P_{2n}. A model for this argument is given in the preceding section, where it is proved that P_2 implies P_4. Again we presume that $a_1 \neq a_2$, and we want to prove that

$$a_1 + a_2 + \cdots + a_{2n} > 2n(a_1 a_2 \cdots a_{2n})^{1/2n}. \tag{5}$$

We know from Theorem 2.2b that

$$a_1 + a_2 > 2\sqrt{a_1 a_2}, \quad a_3 + a_4 \geq 2\sqrt{a_3 a_4},$$
$$a_5 + a_6 = 2\sqrt{a_5 a_6}, \ldots, \quad a_{2n-1} + a_{2n} \geq 2\sqrt{a_{2n-1} a_{2n}}.$$

Adding these, we get the strict inequality

$$a_1 + a_2 + \cdots + a_{2n} > 2\left[\sqrt{a_1 a_2} + \sqrt{a_3 a_4} + \cdots + \sqrt{a_{2n-1} a_{2n}}\,\right]. \tag{6}$$

To the sum of the n terms in square brackets we apply the proposition P_n. These terms might all be equal so we get

$$2[\sqrt{a_1 a_2} + \sqrt{a_3 a_4} + \cdots + \sqrt{a_{2n-1} a_{2n}}] \geq 2n(a_1 a_2 \cdots a_{2n})^{1/2n},$$

not a strict inequality. This with (6) gives the strict inequality

$$a_1 + a_2 + \cdots + a_{2n} > 2n(a_1 a_2 \cdots a_{2n})^{1/2n},$$

and this establishes the proposition P_{2n}.

2.6. Techniques for Finding Extrema. We now apply the inequality of the arithmetic-geometric means to problems in maxima and minima. Algebraic applications are given in this chapter, geometric results in the next.

THEOREM 2.6a. *If n positive functions have a fixed product, their sum is a minimum if it can be arranged that the functions are equal. On the other hand, if n positive functions have a fixed sum, their product is a maximum if it can be arranged that the functions are equal.*

By "positive" functions we mean functions that are positive in the domain or region in which we are interested. The symbol n denotes any positive integer; the theorem does not apply to an infinite set of functions.

Before giving a proof of this result, let us use it in a couple of cases. Consider the problem of minimizing $x^2 + y^2 + z^2$ subject to the restriction that x, y, and z are positive variables satisfying $xyz = k^3$, where k is a positive constant.

(This problem can be stated in a geometric form: Find the shortest distance from the origin to the surface $xyz = k^3$.

The distance from the origin (0, 0, 0) to any point (x, y, z) is $(x^2 + y^2 + z^2)^{1/2}$. To find the shortest distance, it suffices to minimize its square, $x^2 + y^2 + z^2$.)

Solution. Noting that $x^2 + y^2 + z^2$ is a sum of positive functions with a fixed product k^6, the minimum occurs if we can make the terms equal; that is, $x^2 = y^2 = z^2$. These equations along with $xyz = k^3$ give $x = y = z = k$. Hence the minimum value of $x^2 + y^2 + z^2$ is $k^2 + k^2 + k^2$, or $3k^2$.

As a second illustration of the theorem, consider the question of finding the least value of

$$\frac{x}{y} + \frac{3y}{z} + \frac{9z}{x}$$

over all positive real numbers x, y and z.

Solution. Again we have a sum of positive functions whose product is the constant 27. Hence the sum can be minimized if we can solve the equations

$$\frac{x}{y} = \frac{3y}{z} = \frac{9z}{x}.$$

Here we have three equal fractions with product 27; so each of them must equal 3. Thus $x/y = 3y/z = 9z/x = 3$. It follows that the least value of the given sum is $3 + 3 + 3$, or 9. (This minimum occurs for infinitely many values of x, y, and z, because the equations to be solved have infinitely many solutions. We can find actual values of x, y, and z by assigning to one of these variables a positive value in an arbitrary way. Setting $z = 1$, for example, we find $x = 3$ and $y = 1$.)

Now we turn to a proof of Theorem 2.6a. The result is, in fact, a very easy consequence of the inequality of the arithmetic-geometric means. Let $f_1, f_2, \ldots, f_n$ be positive functions having a constant product, say

$$f_1 f_2 \cdots f_n = c.$$

If A and G denote the arithmetic and geometric means of $f_1, f_2, \ldots, f_n$, then $A \geq G$ in the form $nA \geq nG$ can be written as

$$f_1 + f_2 + \cdots + f_n \geq n(f_1 f_2 \cdots f_n)^{1/n} = nc^{1/n}.$$

Thus the value of the sum $f_1 + f_2 + \cdots + f_n$ is at least $nc^{1/n}$, and this minimum is achieved in case $f_1 = f_2 = \cdots = f_n$.

(It is not possible in all problems to satisfy these equations, and that is why the words "if it can be arranged that the functions are equal" appear in the statement of the theorem. This limitation on the procedure is examined later in this section.)

Continuing with the proof, consider next a set of positive functions with a constant sum k, say

$$f_1 + f_2 + \cdots + f_n = k.$$

The arithmetic mean of the functions is therefore k/n. Writing the basic inequality $A \geq G$ in the form $G^n \leq A^n$, we now have

$$G^n = f_1 f_2 \cdots f_n \leq A^n = (k/n)^n.$$

The expression $(k/n)^n$ is a constant, so that these relations can be interpreted as saying that the product $f_1 f_2 \cdots f_n$ has an upper bound $(k/n)^n$. This upper bound is achieved, and so the product is a maximum, if we can arrange that $f_1 = f_2 = \cdots = f_n$.

In the application of this theorem, the functions may need minor adaptations to meet the specifications. Consider, for example, the problem of finding the minimum of $r^4 + s^4 + 2t^2$ over all positive numbers r, s, t satisfying $rst = 81$. Although the terms r^4, s^4, $2t^2$ of the sum do not have a constant product, a slight adjustment will remedy this. If we write the sum as

$$r^4 + s^4 + t^2 + t^2, \tag{1}$$

the terms now have a constant product $r^4 s^4 t^2 t^2 = r^4 s^4 t^4 = 81^4$. Hence we can minimize the sum if the terms of the sum in the form (1) can be made equal. We want therefore to solve the equations

$$r^4 = s^4 = t^2 \quad \text{and} \quad rst = 81$$

in positive numbers. This gives $r^2 = s^2 = t$ and $r = s$ so $rst = 81$ becomes $r^4 = 81$. Thus $r = 3$, $s = 3$, $t = 9$ and the minimum value of $r^4 + s^4 + 2t^2$ is 324.

This example illustrates one way in which a problem needs adjustment in order to meet the specifications of Theorem 2.6a. Other devices are: reformulating a product so that the sum of the factors is a constant; finding the extreme values of a square or cube or some other power of a function instead of the function itself; finding the maximum value of a function by looking for the minimum value of the reciprocal of the function, and of course the dual of this, finding the minimum by looking for the maximum of the reciprocal. We give a series of examples to illustrate these variations.

Example 1. Find the minimum of

$$\frac{12}{x} + \frac{18}{y} + xy$$

over all positive numbers x and y.

Solution. We note that the three terms in the sum have a fixed product $12 \cdot 18$ or 216. The sum is therefore a minimum if we can find x and y to satisfy

$$\frac{12}{x} = \frac{18}{y} = xy.$$

Since the product is 216, or 6^3, each of $12/x$, $18/y$, and xy must equal 6, and hence the required minimum is $6 + 6 + 6$ or 18, occurring iff $x = 2$ and $y = 3$.

Example 2. For any positive constant c, find the maximum value of $xy(c - x - y)$ over all positive numbers x and y.

Solution. There is no point in considering positive numbers x and y so large that $c - x - y < 0$, because then the product $xy(c - x - y)$ is negative and is not a maximum. So we consider values of x and y such that x, y and $c - x - y$ are positive, and we note that their sum is the constant c. Hence we maximize the product if we equate the factors of the product, $x = y = c - x - y$, and this leads easily to the solution $x = y = c/3$. The answer to the question is thus $(c/3)^3$.

As we have seen, it is necessary in some cases to adjust the terms of a sum, or a product, to make the situation conform to Theorem 2.6a.

Example 3. If a is any positive constant, find the minimum value of $x^2 + a/x$ over all positive values of x.

Solution. The product of x^2 and a/x is not a constant; but if we write the sum as

$$x^2 + \frac{a}{2x} + \frac{a}{2x},$$

then the product of the three terms in the sum is the constant $a^2/4$. Hence the sum is minimized if we make these three terms equal, and this leads to $x^2 = a/2x$, $x^3 = a/2$, or $x = (a/2)^{1/3}$. Thus the minimum value of $x^2 + a/x$ is readily seen to be $3(a/2)^{2/3}$.

This example and the earlier one about minimizing $r^4 + s^4 + 2t^2$ suggest the following general principle:

Consider the problem of minimizing a certain sum S, subject to the condition that a related product P is a constant. If the sum can be formulated so that the product of the terms in the sum is P, or some power of P, then the problem can perhaps be solved by use of Theorem 2.6a. The same principle holds for the problem of maximizing a product P, subject to the condition that a certain related sum S is a constant. If the product can be formulated so that the factors in the product have sum S, again it is possible that Theorem 2.6a can be applied.

Example 4. Find the maximum value of the product

$$xy(72 - 3x - 4y)$$

for positive values of x and y.

Solution. The terms of the product x, y and $72 - 3x - 4y$ do not have a constant sum. However, if we adjust the product and write it as

$$\frac{1}{12}(3x)(4y)(72 - 3x - 4y), \tag{2}$$

then ignoring the multiplier 1/12 for a moment we see that the factors $3x$, $4y$, and $72 - 3x - 4y$ have a constant sum 72. Thus we seek values of x and y satisfying

$$3x = 4y = 72 - 3x - 4y.$$

These two equations are easily solved to give the unique values $x = 8$, $y = 6$. Thus the maximum value of the product $xy(72 - 3x - 4y)$ is $8 \cdot 6 \cdot 24 = 1152$.

In examining the product (2) above, we ignored the multiplier 1/12. It may seem obvious, but the observation is worth making that such positive constants can be ignored as multipliers or as added or subtracted constants.

Example 5. Find the smallest value of

$$5x + \frac{16}{x} + 21$$

over positive values of x.

Solution. Although the three terms of the sum, $5x$, $16/x$ and 21, have a constant product, we run into trouble if we try to achieve a minimum value by applying Theorem 2.6a with $n = 3$. The reason for this is that we get

$$5x = \frac{16}{x} = 21,$$

and there is no value of x, positive or otherwise, that satisfies these equations.

However, if we set aside the added constant 21 in the problem, and minimize the sum $5x + 16/x$, the question becomes manageable. For now we apply Theorem 2.6a in the case $n = 2$, noting that $5x$ and $16/x$ have a constant product 80. Hence a minimum is achieved by equating these,

$$5x = 16/x, \quad x^2 = 16/5, \quad x = \sqrt{16/5}.$$

Here we are taking only the positive solution of the equation $x^2 = 16/5$ because of the restriction in the problem. Our conclusion is that the smallest value of $5x + 16/x + 21$ among positive values of x is $21 + 8\sqrt{5}$.

Another standard technique is to work with the square or some other power of a function to be minimized or maximized, if this is simpler.

Example 6. Find the smallest value of $\sqrt{x^2 + y^2}$ among all values of x and y satisfying $3x - y = 20$.

From a geometrical viewpoint, this is the same as asking for the shortest distance, or perpendicular distance, from the origin to the straight line $3x - y = 20$ in analytic geometry.

Solution. Since x^2 and y^2 are positive or zero for all real values of x and y, we may as well find the minimum value of $x^2 + y^2$. Now x and y are related by $3x - y = 20$ or $y = 3x - 20$, and so we can write

$$\begin{aligned} x^2 + y^2 = x^2 + (3x - 20)^2 &= 10x^2 - 120x + 400 \\ &= 10(x^2 - 12x) + 400. \end{aligned}$$

From Theorem 2.2a we know that the minimum of $x^2 - 12x$ is -36, occurring at $x = 6$. It follows that the minimum of $10(x^2 - 12x) + 400$, or of $x^2 + y^2$, is

$$10(-36) + 400 = 40.$$

Thus the smallest value of $\sqrt{x^2 + y^2}$ satisfying the restriction $3x - y = 20$ is $\sqrt{40}$, obtained by taking $x = 6$ and $y = -2$.

One other technique involves switching to the reciprocal of a function. If a function to be minimized assumes only positive values, the problem is equivalent to maximizing the reciprocal of the function, and vice versa.

Example 7. Find the maximum and minimum values, if any, of the function $f(x) = \sqrt{100 + x^2} - x$ over the domain $x \geq 0$.

Solution. The reciprocal is more easily studied:

$$\frac{1}{f(x)} = \frac{1}{\sqrt{100 + x^2} - x} = [\sqrt{100 + x^2} + x]/100.$$

It is virtually obvious that the function $1/f(x)$ increases steadily as x increases from $x = 0$, and becomes indefinitely large along with x. Hence $1/f(x)$ has a minimum at $x = 0$, and no maximum on the domain $x \geq 0$. The answer to the question, then, is that $f(x)$ has a maximum value of 10 at $x = 0$, and no minimum on the domain. Although $f(x)$ has no minimum for positive and zero values of x, the analysis shows that it tends toward zero, but does not reach zero, as x increases indefinitely.

The method used in these problems, although applicable to a wide class of functions, has its limitations. First of all, the procedure applies only if we have a sum of terms with a constant product, or a product whose factors have a constant sum. This in itself is a considerable restriction, but even if one of these conditions is satisfied, the method may fail. For example, consider the problem of finding the minimum value of

$$g(x) = 3x^2 + 3x + 80/x^3$$

over all positive values of x. The three terms of this sum have a constant product 720; so we seek a value of x so that $3x^2 = 3x = 80/x^3$. It is not difficult to see that this system of equations has no solution in real numbers, let alone positive real numbers. (The reader familiar with calculus can solve the problem with relative ease, finding the minimum value of $g(x)$ to be $g(2) = 28$.)

Here is a different example to show that, although the method appears to fail, we can find the extremal value by a slight variation. Consider the problem of finding the least value of the sum

$$x^2 + 4x + 4/x + 1/x^2$$

over positive real numbers x. The functions x^2, $4x$, $4/x$, and $1/x^2$ have a constant product 16, so that Theorem 2.6a suggests that we seek a value of x to satisfy $x^2 = 4x = 4/x = 1/x^2$. These equations have no solution, however. Hence this approach to the problem does not work; but we notice that, if we separate the original sum into $x^2 + 1/x^2$ and $4x + 4/x$, we can apply Theorem 2.6a to each of these expressions separately. The least value of $x^2 + 1/x^2$ is 2, occurring in case $x = 1$, and the minimum of $4x + 4/x$ is 8, also occurring in case $x = 1$. It follows that the minimum value of $x^2 + 4x + 4/x + 1/x^2$ is $2 + 8$, or 10.

This problem also shows that the inequality of the arithmetic-geometric means, like most inequalities in mathematics, does not guarantee a maximum or a minimum in all cases. The inequality $A \geq G$, when applied to the given sum, can be written in the form

$$x^2 + 4x + 4/x + 1/x^2 \geq 4\sqrt[4]{16} = 8. \tag{3}$$

This inequality is correct for positive numbers x, but as we have seen, the minimum value of the sum is 10, not 8. There is no positive number x giving equality in (3).

We give one final example to show that a simple transformation is sometimes helpful.

Example 8. Find the least value of

$$f(x) = \frac{(x + 10)(x + 2)}{x + 1}$$

over positive real numbers x.

Solution. Write y in place of $x + 1$, so that

$$f(x) = f(y - 1) = (y + 9)(y + 1)/y = y + 10 + 9/y.$$

The least value of $y + 9/y$ among positive numbers y is 6, obtained by taking $y = 9/y$, with the solution $y = 3$. The least value of $f(x)$ is therefore 16, occurring in case $x = 2$.

B12. Find the largest value of xyz if x, y, and z are restricted to positive real numbers satisfying (i) $x + y + z = 5$; (ii) $2x + 3y + 4z = 36$. *Suggestion for part* (ii): start by maximizing the product $(2x)(3y)(4z)$.

B13. For any positive constants a, b, c, k find the largest value of xyz over positive real numbers x, y, z satisfying $ax + by + cz = k$.

B14. Find the maximum value of x^2y if x and y are restricted to positive real numbers satisfying $6x + 5y = 45$. *Suggestion*: write the equation in the form $3x + 3x + 5y = 45$.

B15. For positive real numbers x and y such that $x > y$, find the smallest value of

$$x + \frac{8}{y(x - y)}.$$

B16. For any positive constant a, find the maximum of each of the following over all positive x:

$$\frac{x}{x^2 + a}; \quad \frac{x^2}{x^3 + a}; \quad \frac{x}{x^3 + a}.$$

B17. Find the maximum of $x\sqrt{1 - x^2}$ over positive values of x less than 1.

B18. Find the largest value of $2x(12 - x^2)$ over positive numbers x.

B19. Find the least value of $r^2 + rh$ over positive numbers r and h such that r^2h is a constant $c > 0$.

B20. Find the positive number that exceeds its cube by the greatest amount.

B21. Find the positive number whose square exceeds its cube by the greatest amount.

B22. Find the minimum value of

$$\frac{50}{x} + \frac{20}{y} + xy$$

over all positive numbers x and y.

B23. Find the minimum value of

$$\frac{x}{y} + \frac{2y}{z} + \frac{4z}{x} + 12$$

over all positive numbers x, y, z.

B24. Minimize the expression $6x + 24/x^2$ over positive numbers x.

B25. Find the maximum value of $12(xy - 4x - 3y)/x^2y^3$ with x and y positive.

B26. Find the least value of $xy + 2xz + 3yz$ for positive numbers x, y, z, satisfying $xyz = 48$.

B27. Find the smallest value of $x^2 + 12y + 10xy^2$ for positive numbers x and y satisfying $xy = 6$.

B28. Multiply out the product

$$(x+y+z)(\frac{1}{x}+\frac{1}{y}+\frac{1}{z})$$

and deduce that the least value of this product is 9 for positive numbers x, y, z. Hence find the least value of $x^{-1}+y^{-1}+z^{-1}$ over positive numbers x, y, z having a constant sum c.

2.7. The Inequality of the Arithmetic-Harmonic Means. Consider n positive numbers $a_1, a_2, \ldots, a_n$ with arithmetic mean A and geometric mean G defined as usual by

$$A=(a_1+a_2+\cdots+a_n)/n, \quad G=(a_1a_2\cdots a_n)^{1/n}.$$

The *harmonic mean* H of $a_1, a_2, \ldots, a_n$ is defined to be the reciprocal of the arithmetic mean of their reciprocals, that is

$$H=[(a_1^{-1}+a_2^{-1}+\cdots+a_n^{-1})/n]^{-1}.$$

Our purpose first is to prove that $A \geq G \geq H$, with equality in both places here iff $a_1=a_2=\cdots=a_n$. Of course we already know $A \geq G$; so we want to prove $G \geq H$. To do this we apply the inequality of the arithmetic-geometric means to the reciprocals a_1^{-1}, $a_2^{-1}, \ldots, a_n^{-1}$ to get

$$(a_1^{-1}+a_2^{-1}+\cdots+a_n^{-1})/n \geq (a_1^{-1}a_2^{-1}\cdots a_n^{-1})^{1/n}.$$

In view of the meanings of H and G, this inequality can be interpreted as $H^{-1} \geq G^{-1}$, and this gives $G \geq H$. Furthermore, there is equality iff $a_1^{-1}=a_2^{-1}=\cdots=a_n^{-1}$, which is equivalent to $a_1=a_2=\cdots=a_n$.

Thus we have proved $G \geq H$, and so it follows that $A \geq G \geq H$. Now we write the result $A \geq H$ in a form called the *inequality of the arithmetic-harmonic means.*

THEOREM 2.7a. *For any positive numbers* $a_1, a_2, \ldots, a_n$,

$$(a_1^{-1}+a_2^{-1}+\cdots+a_n^{-1})(a_1+a_2+\cdots+a_n) \geq n^2, \qquad (1)$$

with equality iff $a_1=a_2=\cdots=a_n$.

The proof of this follows at once from the inequality $A \geq H$, which can be written as

$$(a_1 + a_2 + \cdots + a_n)/n \geq [(a_1^{-1} + a_2^{-1} + \cdots + a_n^{-1})/n]^{-1}.$$

This can be reformulated as (1) by simple algebra.

Example. If x_1, x_2, x_3, x_4 are positive variables with sum 20, what is the minimum of $x_1^{-1} + x_2^{-1} + x_3^{-1} + x_4^{-1}$?

Solution. By Theorem 2.7a we see that

$$x_1^{-1} + x_2^{-1} + x_3^{-1} + x_4^{-1} \geq \frac{16}{x_1 + x_2 + x_3 + x_4} = \frac{16}{20} = \frac{4}{5}.$$

Thus the answer is 4/5, where this minimum occurs iff $x_1 = x_2 = x_3 = x_4 = 5$.

The inequality of the arithmetic-harmonic means will be used in several places later in this book.

B29. For any real numbers, not necessarily positive, prove that

$$(n-1)(a_1^2 + a_2^2 + \cdots + a_n^2) \geq 2[a_1a_2 + a_1a_3 + \cdots + a_{n-1}a_n],$$

where the $n(n-1)/2$ terms in square brackets include all products a_ia_j for all pairs of integers i, j satisfying $1 \leq i < j \leq n$. Also prove that equality holds iff $a_1 = a_2 = \cdots = a_n$. Suggestion: Add the inequalities $a_i^2 + a_j^2 \geq 2a_ia_j$ for all pairs i, j. For the beginner it might be instructive to write the proof out in detail for (say) $n = 4$.

B30. The *root mean square* of a set of numbers $a_1, a_2, \ldots, a_n$ is another kind of average, defined by

$$R = [(a_1^2 + a_2^2 + \cdots + a_n^2)/n]^{1/2}.$$

Prove that $R \geq A$, where A denotes the arithmetic mean, with equality iff $a_1 = a_2 = \cdots = a_n \geq 0$.

B31. For any real numbers $a_1, a_2, \ldots, a_n$, prove that the minimum value of the function

$$f(x) = (a_1 - x)^2 + (a_2 - x)^2 + \cdots + (a_n - x)^2$$

is $f(A)$, where A is the arithmetic mean of the numbers. (*Remark.* Viewing the numbers $a_1, a_2, \ldots, a_n$ as a set of data, the *variance* of the data is defined as the average $f(A)/n$, and the square root of this is called the *standard deviation.* These are measures of dispersion of the data, because such a term as $(a_1 - A)^2$ is the square of the deviation of a_1 from the average value A. In the language of the preceding problem, the standard deviation is the root mean square of the deviations of the data from the average A.)

B32. The *median* of a set of numbers is the middle one when the numbers are arranged in order of size; or if there is no middle one, the median is the arithmetic mean of the two middle ones. For example, the median of 1, 3, 5, 7, 9 is 5, whereas the median of 1, 3, 5, 7, 9, 11 is 6. Give an example of five positive numbers whose median exceeds the root mean square, and another set of five positive numbers whose median is less than the harmonic mean.

2.8. The Number *e*. The number *e* is a fundamental mathematical constant, comparable to π in importance. It serves as a base for the logarithmic and exponential functions. Although we shall not develop these uses of *e* here, we discuss its basic definition and value as an application of the inequality of the arithmetic-geometric means. The development is so straightforward that parts of it are left to the reader as problems.

For any positive integer n define

$$f(n) = (1 + 1/n)^n \quad \text{and} \quad g(n) = (1 + 1/n)^{n+1}, \tag{1}$$

so that for example $f(1) = 2$, $f(2) = 9/4$, $f(3) = 64/67$, $g(1) = 4$, $g(2) = 27/8$, and $g(3) = 256/81$. First we prove that

$$f(1) < f(2) < f(3) < \cdots < f(n) < f(n + 1) < \cdots. \tag{2}$$

To do this we apply the inequality of the arithmetic-geometric means to the $n + 1$ numbers

$$1, 1 + 1/n, 1 + 1/n, \ldots, 1 + 1/n$$

whose sum is $n + 2$ and whose product is $f(n)$. Thus we get

$$\frac{n + 2}{n + 1} > [f(n)]^{1/(n+1)} \quad \text{or} \quad \left(\frac{n + 2}{n + 1}\right)^{n+1} > f(n).$$

This is the same as $f(n + 1) > f(n)$.

The second result needed is similar, namely, that

$$g(1) > g(2) > g(3) > \cdots > g(n - 1) > g(n) > \cdots. \tag{3}$$

This can be proved by applying the inequality of the arithmetic-geometric means to the $n + 1$ numbers

$$1, 1 - 1/n, 1 - 1/n, 1 - 1/n, \ldots, 1 - 1/n,$$

whose sum is n and whose product is $(1 - 1/n)^n$. Thus we get

$$\frac{n}{n+1} > \left(1 - \frac{1}{n}\right)^{n/(n+1)}.$$

Raising both sides to the power $n + 1$ gives

$$\left(\frac{n}{n+1}\right)^{n+1} > \left(1 - \frac{1}{n}\right)^{n} \quad \text{or}$$

$$\left(\frac{n}{n+1}\right)^{n+1} > \left(\frac{n-1}{n}\right)^{n}.$$

Taking reciprocals we reverse the inequality to get

$$\left(\frac{n+1}{n}\right)^{n+1} < \left(\frac{n}{n-1}\right)^{n}.$$

This is the same as $g(n) < g(n - 1)$, from the definition of $g(n)$ in (1).

Next it is proved that any number in the increasing sequence (2) is less than any number in the decreasing sequence (3), that is,

$$f(m) < g(k) \tag{4}$$

for any positive integers m and k. This is done in three parts.

B33. Prove that $f(n) < g(n)$ for any positive integer n.

This problem establishes (4) in case $m = k$. There are two other cases, $m < k$ and $m > k$, treated separately. In case $m < k$ we can write

$$f(m) < f(k) \quad \text{and} \quad f(k) < g(k),$$

the first from the increasing sequence (2), and the second from problem B33. Taken together, these inequalities give (4) in case $m < k$.

B34. Prove the inequality (4) in case $m > k$.

Now the special case of (4) with $k = 1$ is the inequality $f(m) < g(1)$, and since $g(1) = 4$ we see that

$$f(n) < 4 \quad \text{for all positive integers } n. \tag{5}$$

The next result in the chain is again left to the reader.

B35. Prove that $g(n) - f(n) = f(n)/n$, so that $g(n) - f(n) < 4/n$.

Although $g(n) - f(n)$ is positive for every positive integer n, we see that it can be made as small as we please by taking n sufficiently large, because $4/n$ gets small as n gets large. Thus we have an increasing sequence (2) each of whose terms is smaller than every term of the decreasing sequence (3), but such that the nth terms of the two sequences are very close for large values of n by B35. It is intuitively to be expected under these circumstances that there is a unique number larger than every term in sequence (2), but smaller than every term in sequence (3). (This property is established by a more thorough analysis of the real number system, beyond the scope of this book.) *This number is denoted by the symbol* e, and it is the only number that satisfies

$$f(n) < e < g(n), \quad \text{that is, } (1 + 1/n)^n < e < (1 + 1/n)^{n+1}$$

for all of $n = 1, 2, 3, 4, \ldots$.

It can be calculated that to four decimal places $f(10^4) = 2.7181$ and $g(10^4) = 2.7184$. This shows that e = 2.718 to 3 decimal places of accuracy. To seven decimal places it is 2.7182818. Like π, the number e is irrational and so has an infinite decimal expansion.

Here is a mnemonic device for remembering the first few digits in the decimal expansion of e:

He studied a treatise on calculus

2. 7 1 8 2 8

There are similar devices for the decimal expansion of π, such as this one in Spanish:

¡Sol y Luna y Mundo proclaman al Eterno Autor del Cosmos!

3. 1 4 1 5 9 2 6 5 3 6

The last digit here, "6", is rounded up from the actual value 5 because the next three digits are 8, 9, 7.

2.9. Cauchy's Inequality. Although other inequalities can be used to solve problems in maxima and minima, no result has as many applications as the arithmetic-geometric mean relationship. However, the inequality of Cauchy, which we now develop with applications, has some powerful consequences.

For any two sets of real numbers $a_1, a_2, \ldots, a_n$ *and* $b_1, b_2, \ldots, b_n$*, the inequality*

$$(a_1^2 + a_2^2 + \cdots + a_n^2)(b_1^2 + b_2^2 + \cdots + b_n^2)$$

$$\geq (a_1b_1 + a_2b_2 + \cdots + a_nb_n)^2 \qquad (1)$$

holds, with equality iff the two sets of numbers are proportional.

The two sets of numbers are said to be *proportional* if $a_ib_j = a_jb_i$ for all pairs of subscripts i, j from 1 to n. If none of the numbers is zero, these conditions can be stated in the simple form

$$\frac{a_1}{b_1} = \frac{a_2}{b_2} = \cdots = \frac{a_n}{b_n}, \qquad (2)$$

which is the standard way of indicating proportionality.

We prove Cauchy's inequality (1) in full detail only in the case $n = 3$. The proof generalizes at once to the case of any positive integer n, without difficulty. It is easy to verify the identity

$$(a_1^2 + a_2^2 + a_3^2)(b_1^2 + b_2^2 + b_3^2) - (a_1b_1 + a_2b_2 + a_3b_3)^2 \qquad (3)$$
$$= (a_1b_2 - a_2b_1)^2 + (a_1b_3 - a_3b_1)^2 + (a_2b_3 - a_3b_2)^2,$$

by writing the algebraic expressions in full, and observing that the two sides of the equation are identical after some cancellation. The right member, a sum of three squares, is positive or zero under all circumstances. The inequality (1), in the case $n = 3$, follows at once. Moreover, we see that the equality

$$(a_1^2 + a_2^2 + a_3^2)(b_1^2 + b_2^2 + b_3^2) = (a_1b_1 + a_2b_2 + a_3b_3)^2$$

holds iff

$$a_1b_2 = a_2b_1, \quad a_1b_3 = a_3b_1, \quad \text{and} \quad a_2b_3 = a_3b_2.$$

This proves the Cauchy inequality in case $n = 3$.

In the general case, the identity (3) has the form

$$\begin{aligned}(a_1^2 + a_2^2 + \cdots + a_n^2)(b_1^2 + b_2^2 + \cdots + b_n^2) \\ - (a_1b_1 + a_2b_2 + \cdots + a_nb_n)^2 \\ = (a_1b_2 - a_2b_1)^2 + (a_1b_3 - a_3b_1)^2 + \cdots + (a_{n-1}b_n - a_nb_{n-1})^2,\end{aligned}$$

where the right member is a sum of $n(n-1)/2$ squares. The number of squares here is exactly the number of ways of choosing two distinct subscripts out of the n subscripts $1, 2, \ldots, n$. Once this identity is verified, the inequality (1) follows at once. We do not elaborate on this argument.

Example 1. Find the largest and smallest values of $2x + 3y + 6z$ for values of x, y, z satisfying $x^2 + y^2 + z^2 = 1$. (This is the equation of a sphere of radius 1, center at the origin.)

Solution. By the Cauchy inequality with $n = 3$ and $a_1, a_2, a_3, b_1, b_2, b_3$ replaced by $2, 3, 6, x, y, z$ respectively, we conclude that

$$(2^2 + 3^2 + 6^2)(x^2 + y^2 + z^2) \geq (2x + 3y + 6z)^2, \tag{4}$$

with equality iff

$$\frac{x}{2} = \frac{y}{3} = \frac{z}{6}. \tag{5}$$

On the sphere $x^2 + y^2 + z^2 = 1$, the inequality (4) reduces to

$$49 \geq (2x + 3y + 6z)^2. \tag{6}$$

This inequality assures us that the values of $2x + 3y + 6z$ lie between -7 and $+7$. Furthermore, these *are* the minimum and maximum values we seek, because points do exist on the sphere satisfying the proportionality conditions (5). To verify this, all we have to do is solve the equations (5) with $x^2 + y^2 + z^2 = 1$. The outcome is the pair of solutions

$$x = 2/7, \quad y = 3/7, \quad z = 6/7,$$

and

$$x = -2/7, \quad y = -3/7, \quad z = -6/7.$$

Our conclusion, then, is that the largest value of $2x + 3y + 6z$ on the sphere $x^2 + y^2 + z^2 = 1$ is 7, and the smallest value is -7.

Example 2. Let a, b, c, A, B, C, D be constants such that none of a, b, c is zero and at least one of A, B, C is not zero. Find the minimum of $a^2x^2 + b^2y^2 + c^2z^2$ among values of x, y, z satisfying $Ax + By + Cz = D$.

Remarks on the problem. In the special case $a = b = c = 1$, the question amounts to finding the square of the shortest distance from the origin to the plane $Ax + By + Cz = D$. The condition that at least one of A, B, C is not zero is needed, since otherwise the problem is trivial. The condition that none of a, b, c is zero is not essential; our solution can be easily adapted to other cases.

Solution. We apply the Cauchy inequality in the case $n = 3$ with $a_1, a_2, a_3, b_1, b_2, b_3$ replaced respectively by $ax, by, cz, A/a, B/b, C/c$, to get

$$(a^2x^2 + b^2y^2 + c^2z^2)(A^2/a^2 + B^2/b^2 + C^2/c^2) \geq (Ax + By + Cz)^2.$$

The right member here is the constant D^2, and hence this inequality can be written in the form

$$a^2x^2 + b^2y^2 + c^2z^2 \geq D^2(A^2/a^2 + B^2/b^2 + C^2/c^2)^{-1}, \qquad (7)$$

with equality iff ax, by, cz are proportional to $A/a, B/b, C/c$. It is not difficult to solve these proportionality conditions with $Ax + By + Cz = D$ to get

$$x = b^2c^2AD/k, \quad y = a^2c^2BD/k, \quad z = a^2b^2CD/k,$$

with $k = A^2b^2c^2 + B^2a^2c^2 + C^2a^2b^2$. Thus we have established that equality can be achieved in (7). Our conclusion is that the least value of $a^2x^2 + b^2y^2 + c^2z^2$ on the plane $Ax + By + Cz = D$ is

$$D^2(A^2/a^2 + B^2/b^2 + C^2/c^2)^{-1}.$$

(The name "Cauchy-Schwarz inequality" is often applied to the inequality (1), because Schwarz developed a generalization of this result in integral calculus. The mathematician Buniakowski discovered the same result, independently, so it could be called the Buniakowski inequality.)

Miscellaneous Problems

B36. Find the largest number in the infinite sequence

$$1, \sqrt{2}, \sqrt[3]{3}, \sqrt[4]{4}, \sqrt[5]{5}, \ldots.$$

B37. Find the minimum values of $x^2 + 6x + 1$ and of $x^4 + 6x^2 + 1$ over all real numbers x.

B38. Find the maximum value of $54x^2y^3(1 - x - y)$ over positive numbers x and y.

B39. We know that $x^2 + 1 \geq 2x$ for all positive x. Is it true that $x^3 + 1 \geq 2x$ for all positive x? What is the largest value of the constant k so that $x^3 + 1 \geq kx$ holds for all $x > 0$?

B40. Prove that $a^2/2 + b^3/3 + c^6/6 \geq abc$ holds for all positive a, b, c. Under what circumstances does equality hold?

B41. If a, b, c, d are positive numbers satisfying $a + b = c + d$ and $a^2 + b^2 > c^2 + d^2$, prove that $a^3 + b^3 > c^3 + d^3$.

B42. Find the largest and smallest values, if they exist, of $\sqrt{x^2 + 4x + 85} - \sqrt{x^2 + 4x + 40}$, over all real numbers x.

B43. Let $S(n)$ denote the sum of the digits of the positive integer n. Note that the equation $n = kS(n)$ has a solution n for various values of k: $n = 18$ if $k = 2$, $n = 27$ if $k = 3$, and $n = 12$ if $k = 4$, for example. Find the least positive integer k such that $n = kS(n)$ holds for no integer n.

B44. If a, b, c are real constants, with $a > 0$, determine the value of x so that $ax^2 + bx + c$ is a minimum. *Suggestion:* note that $x^2 - kx$ is a minimum if $x = k/2$ whether k is positive, negative, or zero.

B45. Consider the quadratic polynomial $f(x, y) = ax^2 + 2bxy + cy^2 + dx + ey + k$, where the coefficients are real constants with $a > 0$ and $c > 0$. Observe that in case $b = 0$ the minimum value of $f(x, y)$ can be easily determined by minimizing $ax^2 + dx$ and $cy^2 + ey$ separately. In case $b \neq 0$, verify that the transformation

$$x = X - by/a$$

produces a quadratic polynomial $f(X - by/a, y)$ with no Xy term. Verify also that the coefficients of X^2 and y^2 in this quadratic polynomial are positive iff $a > 0$ and $ac > b^2$, and so $f(x, y)$ has a unique minimum over the domain of all real numbers x and y if these conditions are satisfied. Carry out the details of this process to find the least value of $4x^2 + 16xy + 25y^2 - 24x - 30y + 60$ over all real x and y.

B46. Assume that $abc \neq 0$ in the quadratic polynomial $f(x, y)$ in the preceding problem. Determine necessary and sufficient conditions on the coefficients so that $f(x, y)$ has a minimum, that is, so that there are specific values $x = x_o$, $y = y_o$ such that $f(x_o, y_o) \leq f(x, y)$ for all real x and y.

Notes on Chapter 2

The work of this chapter provides as powerful an application of elementary algebra as can be found. As the French mathematician Jean d'Alembert wrote, "Algebra is generous; she often gives more than is asked of her." The centerpiece of the chapter is the inequality of the arithmetic-geometric means, further applications of which are forthcoming in subsequent chapters. New proofs of this inequality appear in the literature continually, attesting to the appeal of this famous result.

Muirhead (1904) gave a "classification of various proofs" of the famous inequality, along with the sources. Identifying five types of proofs, he pointed out variations that expanded his list to ten. For example, the argument given in §2.3 is the type $B\beta$ in Muirhead's classification. Although many proofs since the time of Muirhead are not very different from those in his listing, new ideas do appear. For example, a different kind of argument is given in the "Postscript on Calculus" following Chapter 12.

Other approaches to the AM-GM inequality can be found in many places, including Beckenbach and Bellman (1961), Kong-Ming Chong (1976), Hardy, Littlewood, and Pólya (1952), Kazarinoff (1961), Rademacher and Toeplitz (1957), and Schaumberger (1977). There are many related inequalities; for example see Klamkin (1968).

§2.6 The algebraic techniques of this section can be pushed beyond what is done here; see Boas and Klamkin (1977).

Algebraic methods can be applied to geometric problems, as shown in the next chapter. Conversely, conclusions in algebra can be drawn from arguments in geometry (in problem D36 at the end of Chapter 4 for example) or in trigonometry (as illustrated in §5.5). Other developments of and variations on these methods can be found in Butchart and Moser (1952),

Fletcher (1971), Frame (1948), Garver (1935), Lennes (1910), Nannini (1967), and Tierney (1953). This is not an exhaustive list.

§2.8. The use of the AM-GM inequality to introduce and define the important constant e was kindly suggested by Peter Lax. For a source, see Mendelsohn (1951).

I am indebted to Robert J. Bitts for the mnemonic device in Spanish for the digits of π. See Pólya (1977, p. 193) for similar epigrams in French and English, including one by the astronomer Eddington, which might be paraphrased roughly as "How I want a drink, chocolate of course, after the later chapters on elusive extremals."

CHAPTER 3

ELEMENTARY GEOMETRIC QUESTIONS

3.1. Introduction. The inequality of the arithmetic-geometric means has many applications to geometry. Most striking perhaps are the so-called isoperimetric results. *Isoperimetric* figures are those having equal perimeters, and among any such class of figures the central problem is to find one with largest area. For example, among all triangles of a specified perimeter, which one has largest area? This question, and the analogous question for quadrilaterals, are discussed in the next two sections. The extension to polygons in general, and to curves in the plane, is given in the next chapter.

Not all the results in the present chapter are of an isoperimetric character, and not all arguments are based on the AM-GM inequality. In §3.5, arguments based on reflection, or mirror images, are used. The best-known result of this type is Heron's theorem, describing the shortest path from a point A to another point B with an intermediate stop at an arbitrary point P on a given line. How to locate the best position for P is the question, as shown in Figure 3.5a. The answer is suggested by the path of a ray of light from point A to point B via a mirror: the light follows the shortest path.

Any plane figure can be reflected in a straight line in the plane, and this reflection is called the *mirror image* of the figure. Similarly, any solid figure can be reflected in a plane. A right-hand glove has a left-hand glove for a mirror image. These simple concepts, used in §3.5, recur at various places later in the book.

3.2. Triangles. We begin with a well-known result about triangles.

THEOREM 3.2a. *Among triangles of a given perimeter, the equilateral triangle has the largest area.*

This is easily established by the algebraic methods of the preceding chapter. Consider any triangle with sides x, y, z such that the perimeter $x + y + z$ is of a specified value. The semi-perimeter s is therefore a constant, $s = (x + y + z)/2$. Maximizing the area amounts to the same thing as maximizing the square

$$A^2 = s(s - x)(s - y)(s - z) \tag{1}$$

from formula (2) of §1.3. Ignoring the constant factor s, we seek the largest value of the product $(s - x)(s - y)(s - z)$. We note that the factors of this product have a constant sum,

$$(s - x) + (s - y) + (s - z) = 3s - x - y - z = s.$$

Hence by Theorem 2.3b the product is a maximum if the factors are equal, $s - x = s - y = s - z$. This gives $x = y = z$ and we have the equilateral triangle.

In the above argument it was necessary to set aside the constant factor s in equation (1). For if we had taken (1) as it stands, we could have noted that the four factors in the product have a constant sum,

$$s + (s - x) + (s - y) + (s - z) = 4s - x - y - z = 2s.$$

However, any attempt to maximize A^2 by making these four factors equal leads nowhere, because the equations $s = s - x = s - y = s - z$ have only the solution $x = y = z = 0$.

The theorem just proved is equivalent to the assertion: among all triangles of a given area, the equilateral triangle has the smallest perimeter. Many isoperimetric results have two formulations like this that are logically equivalent, a maximum area for a fixed perimeter or a minimum perimeter for a specified area. This topic is discussed more fully later in §3.6.

THEOREM 3.2b. *Among all triangles having a specified base and a specified perimeter, the isosceles triangle on that base has the largest area.*

For example, among all triangles with base 16 and perimeter 36, the one with largest area has sides 16, 10, 10, giving area 48.

A proof of this result can be given along the lines of the argument for Theorem 3.2a. That is, if the length of the base is b and the other two sides are denoted by x and y, then the perimeter $b + x + y$ is a specified constant. Thus the semi-perimeter $s = (b + x + y)/2$ is also a constant. The square of the area is

$$A^2 = s(s-b)(s-x)(s-y).$$

To maximize this we ignore the constant factors s and $s - b$, and seek to make $(s - x)(s - y)$ as large as possible. The factors here have a constant sum because

$$(s-x)+(s-y) = 2s - x - y = (b+x+y) - x - y = b.$$

So the maximum occurs if $s - x = s - y$ or $x = y$, as was to be proved.

The proof above is algebraic in character. Any reader familiar with basic properties of the ellipse can easily formulate a geometric proof along the following lines. Denote the vertices of the triangle by P, Q, R, where QR is the base of given length b. Thus Q and R are fixed points, and we consider all possible positions of P (say on one side of QR only for convenience) subject to the condition that $PQ + PR = 2s - b$. Then the point P takes positions along an ellipse with foci at Q and R. Now the area of the triangle PQR, taken in the form one-half the base times the altitude, is a maximum if the altitude is a maximum, the base being fixed. This occurs if P is located at one end of the minor axis of the ellipse, so that $PQ = PR$, as was to be proved.

THEOREM 3.2c. *Among triangles with two sides of specified lengths and the third side of arbitrary length, the triangle of largest area is the right triangle with the arbitrary side as hypotenuse. That is, if the specified lengths are a and b, then the largest triangle has sides a, b, and* $\sqrt{a^2 + b^2}$, *and the area is* $ab/2$.

Let θ denote the angle between the two sides of given lengths a and b. Thus θ is a variable angle, and the problem can be reformulated in this way: choose θ so as to maximize the area. The area is $\frac{1}{2}ab \sin \theta$, so the problem is to maximize $\sin \theta$. The largest value of $\sin \theta$ is 1, occurring iff $\theta = 90°$ for angles in a triangle.

C1. Prove that $A \leq \sqrt{3}L^2/36$ for any triangle, where A is the area and L the perimeter, with equality iff the triangle is equilateral.

C2. Among all right triangles with a given hypotenuse, which one has maximum area? Which one has maximum perimeter?

C3. Given a triangle ABC and a variable point P in the plane of the triangle, prove that $PA^2 + PB^2 + PC^2$ is a minimum if P is located at the centroid of the triangle. *Suggestion:* Use a coordinate system, say with (a_1, a_2), (b_1, b_2), (c_1, c_2) as the coordinates of the vertices, and (x, y) for P. The *centroid* of the triangle has coordinates

$$((a_1 + b_1 + c_1)/3, (a_2 + b_2 + c_2)/3).$$

C4. Among all triangles PQR with a specified angle at P and $PQ + PR = c$, a constant, prove that the one with $PQ = PR$ has (i) maximum area, and (ii) minimum length QR.

3.3. Quadrilaterals. The following problem is given in many calculus books as a basic illustration: among all rectangles of a specified perimeter, which one has the largest area? This is easily solved by the algebraic methods of the preceding chapter. For if the specified perimeter is $2c$, and if x is the length of one side, then $c - x$ is the length of an adjacent side. The area is thus $x(c - x)$ or $cx - x^2$. By Theorem 2.2a the maximum value of $cx - x^2$ is $c^2/4$, occurring iff $x = c/2$. With this value of x we get $c - x = c/2$, so the sides are equal. Thus the square has largest area among rectangles of a given perimeter.

It is natural to put this problem in a more general setting, as in the following result.

THEOREM 3.3a. *Among all quadrilaterals of a specified perimeter, the square has the largest area.*

First we observe that it suffices to consider convex quadrilaterals, because any nonconvex quadrilateral can be replaced by a convex quadrilateral with sides of the same lengths but having a larger area. This is illustrated in Figure 3.3a, where the vertex Q of the nonconvex quadrilateral $PQRS$ lies inside the triangle formed by the other three vertices. Let Q' be the mirror image of the point Q in the line PR, that is, the point such that the line segment QQ' is the perpendicular bisector of the line segment PR. Then the convex

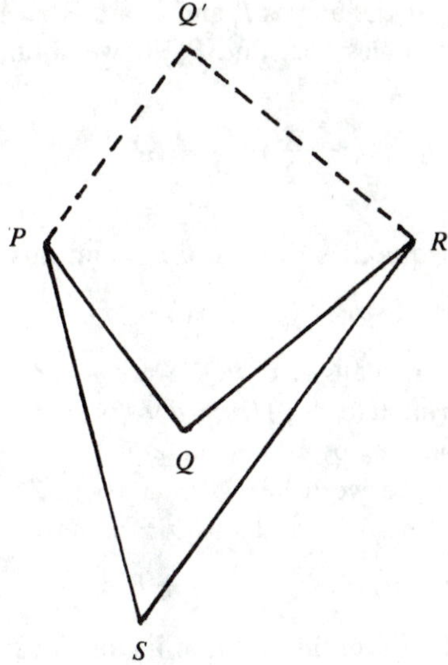

FIG. 3.3a

quadrilateral $PQ'RS$ has sides of the same lengths as $PQRS$, but larger area.

Next, consider any convex quadrilateral $PQRS$ with sides of lengths a, b, c, d in that order, as illustrated in Figure 3.3b. Note that a square of the same perimeter has each side of length $(a + b + c + d)/4$, and area $(a + b + c + d)^2/16$. Denoting the area of $PQRS$ by A, we prove that

$$A \leq (a + b + c + d)^2/16, \tag{1}$$

with equality here iff the quadrilateral $PQRS$ is itself a square.

Let θ denote the angle at the vertex P, between the sides of lengths a and b. Then the triangle PQS has area $\frac{1}{2}ab \sin \theta$, which is at most $\frac{1}{2}ab$, occurring iff $\theta = 90°$. Similarly, the area of the triangle QRS is at most $\frac{1}{2}cd$. It follows that the area A of the quadrilateral satisfies the inequality

$$A \leq \frac{1}{2}ab + \frac{1}{2}cd, \tag{2}$$

with equality here iff the angles P and R are 90°. Applying a similar argument to the triangles PSR and PQR, we obtain

$$A \leq \frac{1}{2}ad + \frac{1}{2}bc. \tag{3}$$

Multiplying the inequalities by 2, adding, and factoring, we get

$$4A \leq (a + c)(b + d),$$

with equality here iff all angles of $PQRS$ are 90°, that is, iff $PQRS$ is a rectangle. (The formula $(a + c)(b + d)/4$ is an ancient Egyptian approximation for the area of a quadrilateral.)

From Theorem 2.2b we have $xy \leq (x + y)^2/4$, with equality iff $x = y$. Replacing x by $a + c$ and y by $b + d$ in this inequality, we get

$$4A \leq (a + c)(b + d) \leq (a + b + c + d)^2/4, \tag{4}$$

and this implies (1). From the chain of inequalities leading to (4), we observe that equality holds here iff $PQRS$ is a rectangle and $a + c = b + d$. These conditions are satisfied iff $PQRS$ is a square, and the proof of the theorem is complete.

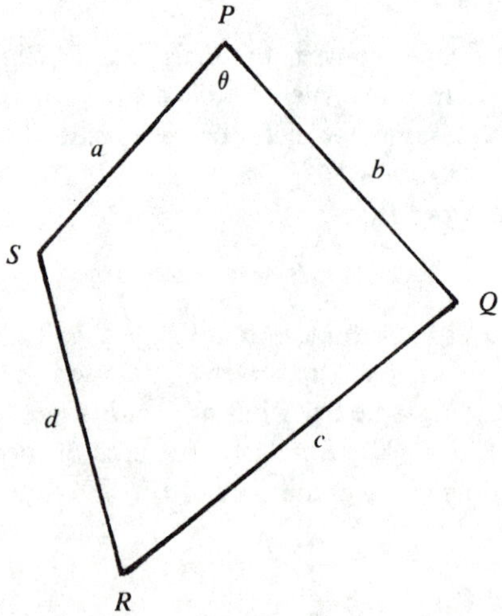

FIG. 3.3b

THEOREM 3.3b. *A quadrilateral inscribed in a circle has a larger area than any other quadrilateral with sides of the same lengths in the same order.*

Let the sides of the quadrilateral be a, b, c, d in that order. The area A of any such quadrilateral was related to the sides, the semi-perimeter s, and a pair of opposite angles θ and λ by the formula

$$A^2 = (s-a)(s-b)(s-c)(s-d) - \frac{1}{2}abcd\,[1+\cos(\theta+\lambda)] \tag{5}$$

in §1.3. The least possible value of $1 + \cos(\theta + \lambda)$ is 0, occurring iff $\theta + \lambda = 180°$, that is, iff the quadrilateral is inscribable in a circle. It follows that A^2 is a maximum in this case, which proves the theorem.

Moreover, formula (5) provides another way to prove Theorem 3.3a. It suffices to consider quadrilaterals having a specified perimeter that are inscribable in some circle. From (5) we see that the area of any such quadrilateral satisfies

$$A^2 = (s-a)(s-b)(s-c)(s-d). \tag{6}$$

Holding the perimeter $2s$ fixed, let a, b, c, d be regarded as variable lengths with a constant sum. Then the sum of the factors in (6) is seen to be constant:

$$(s-a)+(s-b)+(s-c)+(s-d) = 2s.$$

Applying Theorem 2.6a, we find that A^2 is a maximum iff $s - a = s - b = s - c = s - d$, that is, iff $a = b = c = d$. Now the only quadrilateral with equal sides that is inscribable in a circle is a square.

Remark. In the argument above it is tacitly assumed at the outset that corresponding to any quadrilateral Q there *is* a quadrilateral Q' that is inscribable in a circle, having sides of the same lengths as Q. This is not difficult to see from an intuitive standpoint. For suppose that Q is a given quadrilateral $BCDE$. We imagine that Q has sides of thin rods of fixed lengths, loosely attached at the vertices, B, C, D, E, so that the angles are flexible.

Pulling the vertices B and D away from each other, the sum of the angles at C and E increases, and we can make this sum exceed 180°. Similarly, by pulling the vertices C and E away from each other, we can make the sum of the angles at B and D exceed 180°. At some intermediate position the sum of the opposite angles is 180°, and this gives the quadrilateral Q' that is inscribable in a circle.

C5. Consider the quadrilateral Q_1 of maximum area with sides of lengths 8, 9, 12, 19 in that order. Also consider the quadrilateral Q_2 of maximum area with sides 8, 9, 19, 12, in that order. Find the common area. Each of these quadrilaterals is inscribable in a circle. Which circle is larger?

C6. Among quadrilaterals with the perimeter and the length of one side specified, which has maximum area?

C7. Find the largest area among quadrilaterals with sides of lengths 1, 4, 7, 8.

C8. A farmer has 600 feet of fencing to enclose a rectangular region along the bank of a river which can be presumed to be a straight line. If the fencing is to be used along three sides of the rectangle only, with none needed along the river bank, what are the dimensions of the rectangle for maximum area of the rectangle?

C9. Suppose that in the preceding problem the 600 feet of fencing is to be used for the three sides of any quadrilateral, again with no fencing needed along the river. What should be the shape and dimensions of the quadrilateral to maximize the area enclosed? Stated otherwise, among all quadrilaterals with one side of arbitrary length, and the other three sides of total length 600 (or any prescribed positive constant), which has maximum area?

C10. Prove that there does not exist, among all nonconvex quadrilaterals with a specified perimeter c, one having maximum area. Prove that every such quadrilateral has area $<c^2\sqrt{3}/36$, and that there are quadrilaterals of the specified type with area as close to this constant as we please.

C11. Let the lengths of the sides of a convex quadrilateral be a_1, a_2, a_3, a_4, and let the lengths of the diagonals be p_1, p_2. Prove that

$$a_1 + a_2 + a_3 + a_4 < 2p_1 + 2p_2 < 2(a_1 + a_2 + a_3 + a_4).$$

C12. In the preceding question assume that $a_1 \leq a_2 \leq a_3 \leq a_4$ (so that a_1, a_2, a_3, a_4 are not necessarily the lengths of the sides in order around the quadrilateral) and that $p_1 \leq p_2$. Consider the eight inequalities $p_i > a_j$ for $i = 1, 2$ and $j = 1, 2, 3, 4$, and the six inequalities $p_1 + p_2 > a_r + a_s$ with $1 \leq r < s \leq 4$. Which of these inequalities hold for all convex quadrilaterals, and which hold for only some quadrilaterals?

3.4. Miscellaneous Results in Geometry. There are many extremal problems in mathematics that can be solved by the method developed from the inequality of the arithmetic-geometric means. The preceding two sections furnished some examples of this, and now we continue with a variety of geometric problems.

PROBLEM 1. *Given a point P in an angular region QOR as shown in Figure* 3.4a, *how should the points H on OQ and K on OR be chosen so that H, P, K are collinear and the area of the triangle HOK is a minimum?*

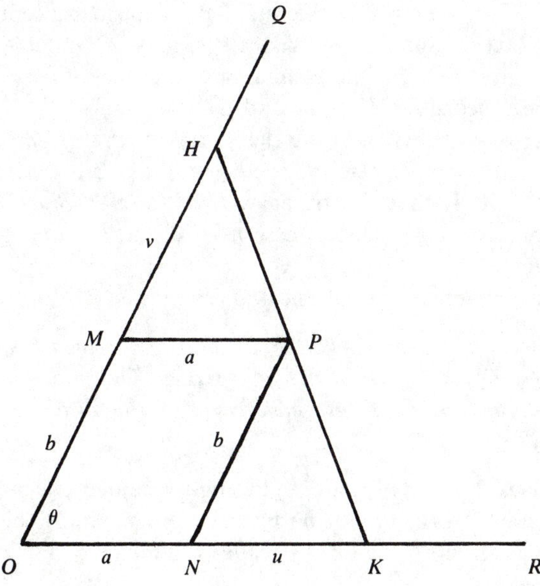

FIG. 3.4a

One way to approach this problem is to draw a line *HPK*, as in Figure 3.4a, and to determine some property or properties of the points *H* and *K* on the assumption that the area of triangle *HOK* is a minimum. Perhaps some such property can be used to locate the correct positions of *H* and *K* to solve the problem. Accordingly, let us draw the line segments *MP* and *PN* parallel to *OR* and *OQ*, respectively, with *M* on *OQ* and *N* on *OR*. The lengths *ON, NK, OM, MH* are denoted by *a, u, b, v*, respectively, so that we also have $MP = a$ and $NP = b$. The area of triangle *HOK* is seen to be $\frac{1}{2}(a + u)(b + v) \sin \theta$, where $\theta = \angle QOR$. Since θ is given, the problem is to minimize $(a + u)(b + v)$, where a and b are constants but u and v are variables depending on the location of the points *H* and *K*. Now u and v are dependent variables because the triangles *HMP* and *PNK* are similar. Corresponding sides of these triangles are proportional, so we see that $v/a = b/u$ or $v = ab/u$. Substituting into $(a + u)(b + v)$, we note that we want to minimize

$$(a + u)(b + ab/u) = 2ab + b(u + a^2/u).$$

Ignoring the constant $2ab$ and the multiplier b, we want to minimize $u + a^2/u$. The two terms in this sum have a constant product; so by Theorem 2.6a we equate u and a^2/u to get $u^2 = a^2$ and $u = a$. Then $v = ab/u$ gives $v = b$. These equations determine unique locations for *H* and *K* because $OK = 2a$ and $OH = 2b$.

Another way of characterizing the solution is that *H and K should be chosen so that P is the midpoint of the line segment HK.* The reason for this is that the triangles *HMP* and *PNK* are not only similar, but also congruent in case $u = a$ and $v = b$, and hence $HP = PK$.

Here is a closely related classical question:

PROBLEM 2. *Given a point P in an angular region QOR with* $\angle QOR = 90°$, *how should the points H on OQ and K on OR be chosen so that H, P, K are collinear and the distance HK is a minimum?*

Attention is restricted to the right angle case here, because otherwise the problem is beyond the reach of our methods. Figure 3.4a can serve as an illustration if the reader will imagine that $\angle QOR = 90°$.

(This problem is sometimes stated in the following way: find the length of the longest ladder that can be moved around a right-angle corner from a corridor of width a to a corridor of width b, as illustrated in Figure 3.4b.)

The problem of minimizing the length HK is the same as minimizing $KP + PH$, or

$$(b^2 + u^2)^{1/2} + (a^2 + v^2)^{1/2}, \quad \text{or}$$

$$(b^2 + u^2)^{1/2} + (a^2 + a^2b^2/u^2)^{1/2},$$

since the relation $v = ab/u$ still holds. This can be written

$$(b^2 + u^2)^{1/2}[1 + a/u]$$

by simple algebra. We minimize the square of this positive function of u, which can be written in the form

$$(a^2 + b^2) + \left[u^2 + \frac{ab^2}{u} + \frac{ab^2}{u}\right] + \left[au + au + \frac{a^2b^2}{u^2}\right].$$

Writing this as $(a^2 + b^2) + f(u) + g(u)$ for convenience, where $f(u)$ and $g(u)$ denote the expressions in the square brackets, we observe that the minimum of $f(u)$ occurs for the same value of u as the minimum of $g(u)$. This is because the three terms in $f(u)$ have a

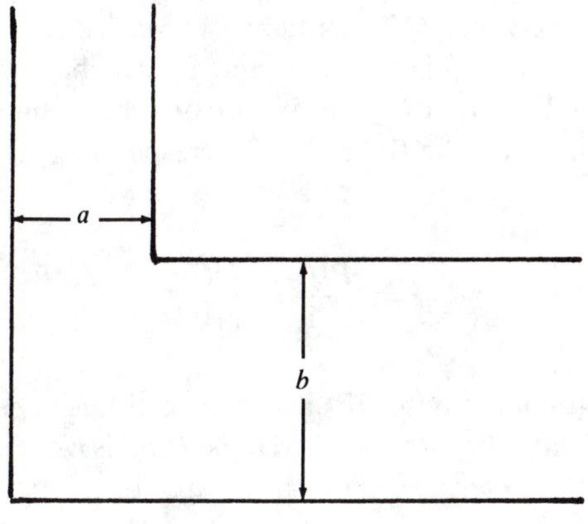

FIG. 3.4b

constant product, and likewise for $g(u)$, so the minimum occurs for $f(u)$ in case $u^2 = ab^2/u$, and for $g(u)$ in case $au = a^2b^2/u^2$. In both cases this gives $u^3 = ab^2$ or $u = (ab^2)^{1/3}$. Then $v = ab/u$ implies $v = (a^2b)^{1/3}$, and the problem is solved.

A natural extension of Problems 1 and 2 above is to ask for the location of the points H on OQ and K on OR, with H, P, K collinear, such that the *perimeter* of the triangle HOK is a minimum. This problem is solved by the auxiliary circle technique in §3.7. Another variation of the problem is to ask for the location of points H and K so as to minimize the product $PH \cdot PK$. This is problem C40 at the end of the chapter.

Another classical problem is the question of the rectangle of largest area that can be inscribed in a given triangle. It is no more difficult to solve this problem in a slightly more general form:

PROBLEM 3. *Prove that in any triangle it is possible to inscribe a parallelogram (and also a rectangle) whose area is half that of the triangle. Prove also that no parallelogram of larger area can be inscribed.*

This result is proved here for inscribed parallelograms all of whose vertices lie on the sides of the triangle. Some remarks about the other possibilities are made at the end of the proof.

Consider any triangle PQR and any inscribed parallelogram whose vertices are on the sides of the triangle. It must be that one side of the triangle, say QR, contains two vertices of the parallelogram as illustrated in Figures 3.4c and 3.4d, where the parallelograms are $QBCD$ and $ABCD$. In both cases the triangles PDC and PQR are similar. Let r be the ratio of corresponding lengths, for example

$$r = \frac{PD}{PQ} = \frac{DC}{QR}.$$

If b is the length of the base QR, and h the altitude of triangle PQR, then $DC = rb$ and the altitude of triangle PDC is rh. The altitude of the parallelogram is $h - rh$, and so the area of the parallelogram is

$$rb(h - rh) = bh(r - r^2).$$

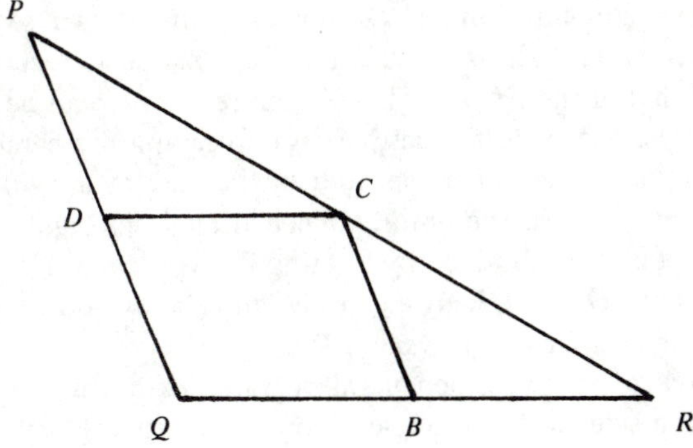

FIG. 3.4c

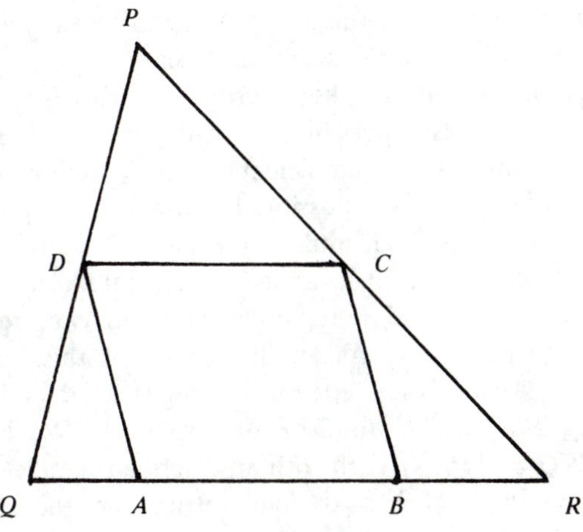

FIG. 3.4d

By Theorem 2.2a the maximum value of $r - r^2$ is $1/4$, occurring in case $r = 1/2$. Thus the largest possible parallelogram has area $bh/4$, which is half the area of triangle PQR.

It is easy to see that such a parallelogram can always be inscribed in a triangle, by choosing D and C as the midpoints of the sides PQ and PR as in Figures 3.4c and 3.4d. In Figure 3.4d, if D and C are chosen as midpoints, and if DA and CB are drawn so as to be perpendicular to QR, we would have an inscribed rectangle with area half that of the triangle. Three such rectangles can be inscribed in a triangle, except in the case of an obtuse-angled triangle, which allows only one such rectangle, and in the case of a right triangle, which allows two. In the obtuse-angled triangle in Figure 3.4c, the unique rectangle is obtained by locating two vertices at D and B, the midpoints of PQ and QR, respectively, and the two other vertices in the obvious positions on the side PR.

The area of any inscribed parallelogram *not* having all four vertices on the sides of the triangle is strictly less than half the area of the triangle. The proof of this is easily obtained, by examining first parallelograms that have exactly three vertices lying on the sides of the triangle, then parallelograms with exactly two, exactly one, and finally with no vertices on the sides. As to the ideas needed in the proofs, the following questions can be raised.

Is it possible to enlarge the inscribed parallelogram to a type already considered? Is it possible to shrink the triangle to a smaller triangle that contains the parallelogram, again reducing the problem to an earlier case? Is it possible to draw a line segment from a vertex of the triangle to an interior point of the opposite side, this line segment being parallel to a side of the parallelogram, thereby splitting the triangle and the parallelogram in two, reducing the problem to earlier cases? As an illustration of this, consider the parallelogram $ABCD$ inscribed in the triangle PQR in Figure 3.4e. By drawing PK parallel to DA, the triangle PQR is split into triangles PKQ and PKR, each with an inscribed parallelogram. The parallelogram in PKQ has all four vertices on the sides of the triangle. As to PKR, if we draw the line segment KB and produce it to meet CR, say at T, then triangle TKP has an inscribed parallelogram with all four vertices lying on the sides of the triangle. All further details of the argument are straightforward, and are left to the reader.

PROBLEM 4. *Find the shortest distance from the point* $(c, 0)$ *to* $y^2 = 4x$, *where* c *may be positive, negative, or zero.*

If (x, y) is any point on the curve the distance to be minimized is

$$\sqrt{(x-c)^2 + (y-0)^2} = \sqrt{(x-c)^2 + 4x}, \tag{1}$$

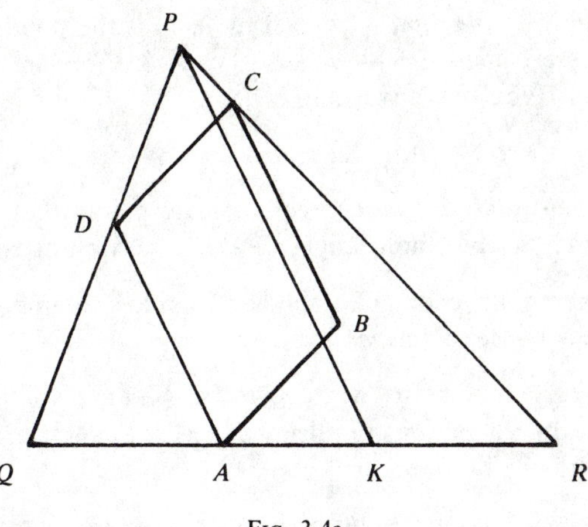

FIG. 3.4e

where y^2 has been replaced by $4x$ from the equation of the curve. To minimize this we look at its square

$$(x - c)^2 + 4x = x^2 + (4 - 2c)x + c^2. \qquad (2)$$

Ignoring the c^2 we seek to minimize $x^2 + (4 - 2c)x$. According to Theorem 2.2a, the minimum here occurs at $x = -(4 - 2c)/2 = c - 2$.

However, at this juncture we must be careful not to let the algebra mislead us. The only values of x that can be used in the coordinates (x, y) on the curve $y^2 = 4x$ are those with $x \geq 0$. So the solution $x = c - 2$ above is valid if $c \geq 2$, leading to the answer $\sqrt{4c - 4}$ in (1).

If $c < 2$ the answer $x = c - 2$ is of no use, so we look again at (2) and seek the minimum value of $x^2 + (4 - 2c)x$ among positive and zero values of x only. Now $4 - 2c$ is positive for $c < 2$, so for positive values of x it is clear that $x^2 + (4 - 2c)x$ is positive. Hence it follows that the least value of $x^2 + (4 - 2c)x$ occurs at $x = 0$. Thus for values of $c < 2$ the shortest distance is obtained from (1) with $x = 0$, namely

$$\sqrt{(-c)^2} = |c|.$$

This is c if $c > 0$, but is $-c$ if $c < 0$.

The geometric interpretation of all this is the following. If $c < 2$, the point on $y^2 = 4x$ that is nearest to $(c, 0)$ is the point $(0, 0)$. It is also the nearest point if $c = 2$. But, if $c > 2$, there are two nearest points to $(c, 0)$ on the curve, namely,

$$(c - 2, \pm 2\sqrt{c - 2}).$$

C13. Given a circle with center C, locate points P and Q on the circumference so that the triangle CPQ has maximum area.

C14. Among all rectangular solids of a fixed volume, prove that the cube has the least surface area.

C15. A rectangular box with no top is to have volume k cubic units. Find the dimensions requiring the least amount of material in the sides and the bottom.

C16. A heated storage building in the shape of a rectangular solid is to be constructed in a cold climate, with volume 75,000 cubic feet. If the heat loss per square foot through the floor is only one-fifth of the loss through the side walls and the roof, what should be the dimensions for minimum loss of heat?

C17. A rectangular box is to lie in the first octant, with one corner at the origin $(0, 0, 0)$ of the coordinate system, and the diagonally opposite corner of the box on the plane $x/a + y/b + z/c = 1$, where a, b, c are positive constants. Find the maximum possible volume of such a box. (The first octant is the region in 3-space where the coordinates (x, y, z) are nonnegative. Thus the problem amounts to maximizing the product xyz subject to $x/a + y/b + z/c = 1$.)

C18. Find the dimensions of the box of maximum volume which can be fitted into the ellipsoid $x^2/a^2 + y^2/b^2 + z^2/c^2 = 1$, assuming that each edge of the box is parallel to one of the axes of the coordinate system. *Suggestion:* The problem amounts to maximizing $8xyz$ subject to the condition $x^2/a^2 + y^2/b^2 + z^2/c^2 = 1$.

C19. Determine the rectangle of greatest area that can be inscribed in a semicircle, with one side of the rectangle lying along the diameter. *Suggestion:* Use a coordinate system, taking the semicircle to be the part of the circle $x^2 + y^2 = r^2$ with $y \geq 0$.

C20. A square piece of sheet metal, of dimensions 30 by 30, is to be used to construct a rectangular box with no top by cutting out square corners and folding up the sides. Find the largest possible volume for such a box.

C21. Among all right circular cylinders of a specified volume, prove that the one having least surface area has $h = 2r$, where h is the height and r the radius of the cylinder. (A *right circular cylinder* is a cylinder whose cross-section is a circle, and whose base is perpendicular to the axis of the cylinder.)

C22. Consider a hollow circular cylinder of a specified volume, with no top. What should be the relationship between h and r (see the preceding problem) for minimum surface area?

C23. Find the relationship between h and r for the right circular cylinder of maximum volume that can be inscribed in a sphere.

C24. Given that the strength of a beam of rectangular cross-section of base b and height h is proportional to bh^2, find the dimensions of the strongest such beam that can be cut out of a circular log of diameter 12 inches by trimming the circular cross-section down to a rectangle. (A two-by-four beam is stronger with base $b = 2$ and $h = 4$ than with base $b = 4$ and $h = 2$. The problem amounts to maximizing bh^2 subject to $b^2 + h^2 = 144$.)

C25. Find the point or points on the curve $y = x^2$ that are closest to $(0, c)$, where c may be positive, negative, or zero.

C26. Points A, B, C are to be selected on the three axes of a rectangular coordinate system, where O is the origin, in such a way that the sum of the lengths of the edges of the tetrahedron $OABC$ is a constant. How should the points be located so as to maximize the volume of the tetrahedron? *Suggestion:* If the lengths OA, OB, OC are denoted by x, y, z, then AB, AC, and BC have lengths

$$\sqrt{x^2 + y^2}, \quad \sqrt{x^2 + z^2}, \quad \text{and} \quad \sqrt{y^2 + z^2}.$$

The volume is $xyz/6$, from the formula $\frac{1}{3}hb$, where b is the area of the base and h is the altitude. Taking triangle OAB as the base, we have $b = xy/2$ and $h = z$.

3.5. The Reflection Principle. Consider the following question: *Given two points A and B on the same side of a line CD, find the*

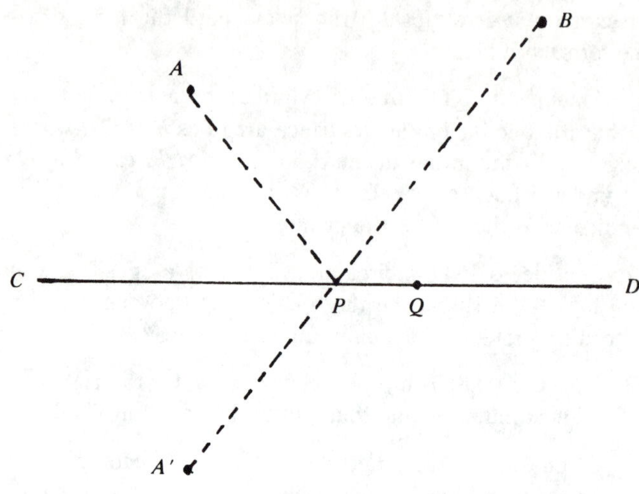

FIG. 3.5a

point P on CD so that AP + PB is a minimum. The quickest way to solve this problem is to take the mirror image in the line CD of one of the points A and B, for example the mirror image A' of A as shown in Figure 3.5a. Then the line BA' intersects CD at the desired point P. This is Heron's theorem.

The concept of a mirror-image point is almost self-explanatory: A' is the point such that the line segment AA' is perpendicular to CD and bisected by CD. To prove that P is the point we want, consider any other point Q on CD as shown in Figure 3.5a. Then we note that $AQ = A'Q$ and $AP = A'P$; hence we get

$$AQ + QB = A'Q + QB > A'B = A'P + PB = AP + PB,$$

by using the triangle inequality.

We can apply this geometric result to algebra by reformulating the problem in a coordinate setting. Let CD be the x axis, with the origin at C, and let A and B have coordinates (a, b) and (c, d) with $b > 0$ and $d > 0$. Denoting any point on CD by $(x, 0)$, the problem is to minimize the sum of the distances from $(x, 0)$ to (a, b) and to (c, d). That is, find the value of x that makes

$$\sqrt{(x-a)^2 + b^2} + \sqrt{(x-c)^2 + d^2} \qquad (1)$$

a minimum. Using the geometric result, the algebra is straightforward. The coordinates of A' are $(a, -b)$, so the equation of the line $A'B$ is

$$\frac{y+b}{-b-d} = \frac{x-a}{a-c}.$$

Setting $y = 0$ and solving for x, we find the x coordinate of the point P where $A'B$ crosses the x axis, namely,

$$x = \frac{ad+bc}{b+d}. \tag{2}$$

This is the unique value of x that makes (1) a minimum.

As another illustration of the reflection principle, consider the problem C8, now stated in a purer geometric form: *Among all rectangles with one side of arbitrary length and the other three sides of constant sum c, which has the largest area?* Here is a solution by the reflection method. Suppose that the answer is the rectangle $PQRS$ as shown in Figure 3.5b, where PS is the side of arbitrary length, and $PQ + QR + RS = c$. Taking the mirror image or reflection of this rectangle in the line PS we get the rectangle $PQ'R'S$. Ignoring the

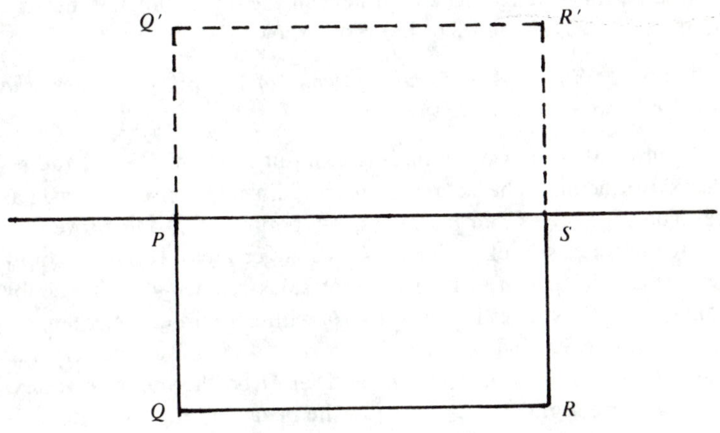

FIG. 3.5b

arbitrary length PS, it must be that the rectangle $QRR'Q'$ has the largest area among all rectangles of perimeter $2c$. But the answer to this is the square, so $PQRS$ must be half a square, with

$$PQ = RS = c/4, \qquad QR = c/2.$$

In problem C8 the value of c is 600, so the answer to that question is a rectangle of dimensions 150 by 300 feet.

Problems C14 and C15 provide another application of the reflection principle, which we now use to solve C15, assuming that we know the solution of C14. By reflecting the entire box as described in C15 in its four top vertices, we must have the entire rectangular solid, the cube, that is, as in C14. Hence the solution to C15 is half a cube. Since the volume is k cubic units, we can take the length, width, and height as x, x, $x/2$ to get the volume equation $x^3/2 = k$, or $x = \sqrt[3]{2k}$. The dimensions are then seen to be as given in the solution of problem C15.

As a final example, if we take problem C21 as solved, we can use it to solve C22 by the reflection principle. For if we reflect the cylinder described in C22 in its open circular top, we get the entire cylinder of C21. Thus from the answer $h = 2r$ in C21 we get the height-radius relationship $h = r$ in C22.

In these examples we have been arguing from the known situation to the half-case, from the square to the half-square, the cube to the half-cube, for instance. The argument in the other direction needs a little more care, as the following result shows.

THEOREM 3.5a. *Among all hexagons of a given perimeter, the regular hexagon has the largest area.*

To prove this we use the half-hexagon of problem C9 and the reflection principle. The idea is to begin with any convex nonregular hexagon H_1 of the given perimeter, say perimeter k, and prove that the regular hexagon of perimeter k has larger area. To H_1 we apply Theorem 3.2b to consecutive pairs of sides replacing each pair by equal sides; thus H_1 is replaced by H_2 with sides in order of lengths a, a, b, b, c, c, say, where $2a + 2b + 2c = k$. Let P be the vertex at the join of two sides of length a, and let Q be the opposite vertex. Then the line segment PQ separates the hexagon into two quadrilaterals with sides a, b. b, PQ in one case, and a, c, c, PQ in the other.

We now reverse one of these quadrilaterals on PQ as a base; to be more specific, we take the mirror image of one of these quadrilaterals in the perpendicular bisector of the line segment PQ, leaving the other quadrilateral unchanged. Now we have a hexagon H_3 with sides in order of lengths $a, b, b, a, c, c,$ as illustrated in Figure 3.5c. In H_3 let R be the vertex at the join of the two sides of length b, and S the opposite vertex, at the join of the two sides of length c. The line segment RS separates the hexagon H_3 into two quadrilaterals having the same perimeter $RS + a + b + c$.

To these quadrilaterals we apply the solution of problem C9, namely, half of a regular hexagon, using RS as the side of arbitrary length. The other three sides in each quadrilateral are replaced by three lengths $(a + b + c)/3$, so that we get two half-hexagons in contiguous positions to form a regular hexagon with sides of length $(a + b + c)/3 = k/6$. The steps from hexagon H_1 to the regular hexagon have involved an increase in area, and so the proof is complete.

3.6. Equivalent Results. We know that among all triangles of a given perimeter c, the equilateral triangle has largest area. This result, Theorem 3.2a, is logically equivalent to the following proposition:

Among all triangles of a given area k, the equilateral triangle has the shortest perimeter.

To say that the two results are logically equivalent means that each implies the other.

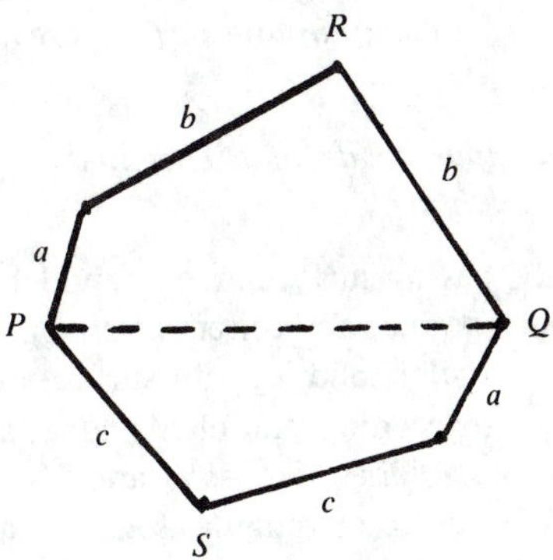

Fig. 3.5c: The Hexagon H_3

Let us use Theorem 3.2a to prove this second formulation. Among all triangles of area k, first consider the equilateral triangle T_1, say with perimeter p_1. (There is, of course, a simple relation between the area and the perimeter of an equilateral triangle, but we do not want to use it here. The point is that the logical argument we are using is not dependent on any such formula. In adaptations of the argument from triangles to other situations, the formulas may be difficult to determine. So we stay with the simple logic of the situation.) We want to prove that there is no other triangle with area k that has a perimeter $\leq p_1$. Suppose, on the contrary, that there is a nonequilateral triangle T_2 of area k having perimeter $p_2 \leq p_1$.

We treat two cases, $p_2 = p_1$ and $p_2 < p_1$. In case $p_2 = p_1$ we have two triangles T_1 and T_2, one equilateral and one nonequilateral, with the same perimeter p_1 and the same area k. This is impossible because the equilateral triangle T_1 is unique in having the largest area among all triangles of perimeter p_1, by Theorem 3.2a.

Second, suppose that $p_2 < p_1$. Let T_3 be the equilateral triangle of perimeter p_2. Let k_3 be the area of T_3. Since T_3 and T_2 have the same perimeter p_2, we conclude that $k_3 > k$ because T_3 is equilateral and T_2 is not. On the other hand, if we compare the equilateral triangles T_1 and T_3, we see that T_3 has the smaller perimeter and hence the smaller area. Thus $k_3 < k$, contradicting the earlier conclusion. Hence our supposition $p_2 \leq p_1$ is false, and this proves that $p_2 > p_1$.

As a second example of equivalent assertions, let us assume the result of Problem C14, *"Among all rectangular solids of a fixed volume, the cube has the minimum surface area,"* and prove the equivalent formulation:

Among all rectangular solids of a specified surface area c, the cube has the largest volume.

Let R_1 be the cube having surface area c, and let its volume be V_1. As in the preceding example, an indirect argument is used. Suppose there is another rectangular solid R_2 with surface area c, and having volume V_2 equal to or exceeding that of the cube, so that $V_2 \geq V_1$.

We treat the two possibilities, $V_2 = V_1$ and $V_2 > V_1$, separately. In case $V_2 = V_1$ we have two rectangular solids R_1 and R_2, the first a cube and the second not a cube, with equal volumes and equal surface areas. This is impossible by problem C14.

Next suppose that $V_2 > V_1$. Then let R_3 be the cube of volume V_2, say with surface area S_3. Comparing R_2 and R_3 we have two solids with the same volume; by problem C14 the cube R_3 has the smaller surface area, so $S_3 < c$. Now if we compare the two cubes R_1 and R_3, this inequality $S_3 < c$ says that R_3 has the smaller surface area, but $V_2 > V_1$ says that R_1 has the larger volume, a contradiction. This completes the proof.

Clearly the proofs in these two examples are virtually identical in form. In general a "constant perimeter, maximum area" result has an equivalent analog with a "constant area, minimum perimeter" form. Similarly, in 3-space we can usually get from a "constant surface area, maximum volume" result an equivalent "constant volume, minimum surface area" form. Henceforth we shall assume the equivalence of such results without proof.

As another example, here is an equivalent formulation of Theorem 3.2b: *Among all triangles with specified area and one fixed side, the triangle with the other two sides equal has minimum perimeter.*

It is nice to get two results for the price of one. But equally important is the fact that this equivalence of results enables us to convert problems to simpler forms in many cases. For example, consider the question:

I. *Among all triangles PQR having a given angle P and a given perimeter, which has maximum area?*

This is equivalent to:

II. *Among all triangles PQR having a fixed angle P and a given area, which has minimum perimeter?*

The second form is easier to deal with, and so we solve it rather than the first. Let x, y, z denote the lengths of the sides PQ, PR, QR, respectively, and let θ be the fixed angle at P. The area A and the perimeter L are given by

$$A = \frac{1}{2}xy \sin \theta, \quad L = x + y + z$$

$$\text{or} \quad L = x + y + (x^2 + y^2 - 2xy \cos \theta)^{1/2}. \tag{1}$$

Now θ and A are constants, so the product xy is also a constant. From Theorem 2.6a we know that $x + y$ is a minimum if $x = y$. Because x^2y^2 is also a constant, we know that $x^2 + y^2$ is a minimum if $x = y$. Looking at the final equation (1), we see that since the term $2xy \cos \theta$ is a constant, L is a minimum also in the case $x = y$.

Therefore, the solution of problem II above, and hence also of problem I, is to require that $x = y$, that is $PQ = PR$, making the triangle PQR isosceles.

3.7. Auxiliary Circles. An auxiliary circle, not necessarily mentioned in a problem, is sometimes of use in finding a solution. For example, consider the following natural variation on the first two problems in § 3.4.

PROBLEM 1. *Given a point P in an angular region QOR, how should the points H on OQ and K on OR be chosen so that H, P, K are collinear and the perimeter of the triangle HOK is a minimum?*

A key to the solution of this problem is to draw a circle C tangent to the lines OQ and OR, passing through the point P. There are two circles meeting these specifications; we want the larger one, as in Figure 3.7a. The solution to the problem is the line HPK that is tangent to this circle at P.

To prove this, let M and N denote the points of tangency of the

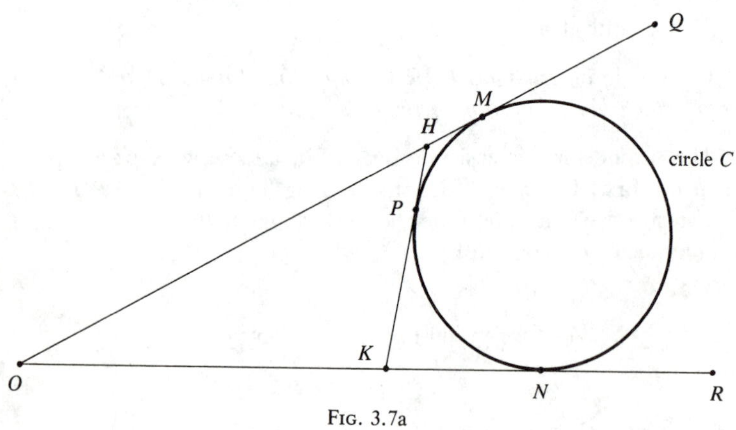

FIG. 3.7a

lines OQ and OR with the circle C. Then the perimeter of the triangle HOK can be written as

$$\begin{aligned} OH + OK + HK &= OH + OK + PH + PK \\ &= OH + OK + HM + KN \\ &= OM + ON = 2(OM), \end{aligned}$$

where we have used the fact that the two tangent lines drawn to a circle from an exterior point are equal in length.

Out of this calculation we can make the following observation: the perimeter of a triangle HOK equals twice the length of a tangent line from O to the circle (lying outside the triangle) that is tangent to HK and to the extensions of the lines OH and OK beyond H and K.

Now consider another line segment H_1PK_1 with H_1 on OQ and K_1 on OR. We want to prove that the perimeter of triangle H_1OK_1 exceeds that of triangle HOK. Using the observation of the preceding paragraph, it suffices to establish that the circle C_1 is larger than the circle C, where C_1 is the circle tangent to the lines H_1K_1, H_1Q, and K_1R. The circle C_1 is larger than C because P lies outside C_1.

Another question whose solution involves an auxiliary circle is the following.

PROBLEM 2. (The picture on the wall.) *A picture hangs on a wall above the level of an observer's eye. How far from the picture should the observer stand to maximize the angle of observation, that is, the angle subtended at the observer's eye by the top and bottom of the picture?*

The problem is illustrated in Figure 3.7b, where Q and R represent the bottom and top of the picture and TS is the horizontal line at the level of the eye of the observer. The problem is to locate the point P on TS so as to maximize the angle QPR. First we solve the problem geometrically, and then give an algebraic formulation of the solution.

A key to the solution is the circle passing through the points Q and R, tangent to the line TS. The point P should be located at the point of tangency, as shown in Figure 3.7c. If K is any other point on TS, we prove that $\angle QPR > \angle QKR$. *Let* H be the intersection point of the line segment RK and the circle, and note that $\angle QPR = \angle QHR$

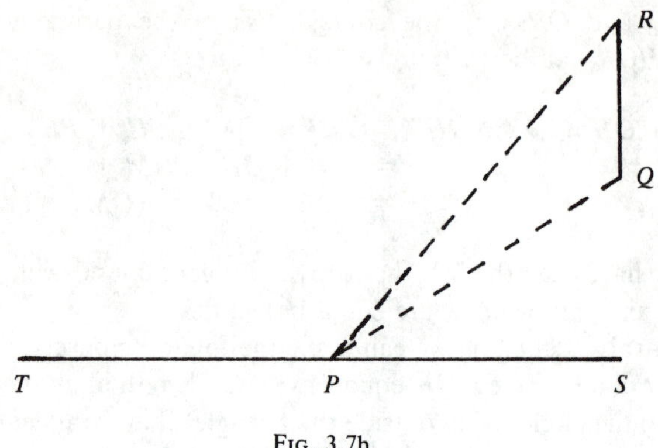

FIG. 3.7b

by the basic property that the chord QR subtends equal angles at any two points on the arc QR. Hence we can write

$$\angle QPR = \angle QHR = \angle QKR + \angle HQK > \angle QKR.$$

Thus we have proved that the angle of observation is maximized if the observer's eye is at the point of tangency P. This solution is valid even in the case where the picture is not flat against the wall, but, as is so often the case, with the picture tilted slightly out from the wall from bottom to top. Even if the line QR is not vertical, there is a unique circle passing through Q and R tangent to the horizontal line at the level of the observer's eye.

This problem is given in many calculus books in the case where the picture is flat against the wall, so that QR is a vertical line segment. The question is usually asked in terms of how far the observer should stand from the plane of the picture, that is, the distance PS in Figure 3.7c. Also, the information is given that the bottom of the picture is b feet above, and the top is c feet above, the level of the eye of the observer. In terms of Figure 3.7c, this means that $QS = b$ and $RS = c$. We now determine PS in terms of b and c.

It is a basic result from the geometry of the circle that the tangent line SP and the secant line SQR satisfy the distance relation

$$SP^2 = SQ \cdot SR.$$

Assuming this, we get the solution

$$SP^2 = bc, \qquad SP = \sqrt{bc}.$$

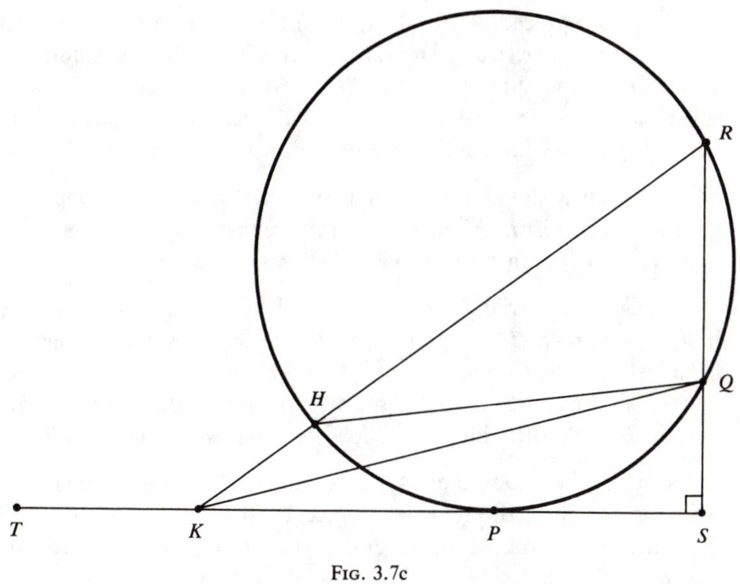

FIG. 3.7c

Miscellaneous Problems

C27. Given an angle $PQR < 180°$ and a positive constant c, how should the points H and K be chosen on the half-lines QP and QR so that $HK = c$ and the triangle HQK has maximum area?

C28. Given two points A and B in the same plane as a line PQ, locate the point K on PQ so that $AK^2 + KB^2$ is a minimum.

C29. (i) Given 3 points, $A, B, C,$ in that order on a straight line, locate the point P on the line so that $PA + PB + PC$ is a minimum. (ii) Given points A, B, C, D on a line in that order, locate P on the line to minimize $PA + PB + PC + PD$. (iii) Generalize to n points on a line, locating P so that the sum of the n distances from P to the points is a minimum.

C30. A room is rectangular in shape, 18 feet long, 14 feet wide, and 10 feet high. A mosquito flies along the shortest path from one corner of the room to the opposite corner, for example from the southeast corner at the floor to the northwest corner at the ceiling. How far does it fly?

C31. In the preceding problem, an ant walks from one corner to the opposite corner, using any parts of the floor, the walls and the ceiling to get the shortest path. (Assume that the floor, walls and ceiling are smooth plane surfaces without indentations or irregularities of any kind.) How far does it walk?

C32. Given a closed straight line segment, how should n points (not necessarily distinct) be chosen on the segment so that the sum of all distances between pairs of points is a maximum?

C33. Given any triangle T with sides of lengths a, b, c, is it true that any triangle T' with shorter sides also has smaller area? Specifically, if the sides of T' are a', b', c' with $a' < a$, $b' < b$, $c' < c$, is area $T' <$ area T? If so, prove it. If not, determine conditions on T so that the inequality does hold for every triangle T'.

C34. Find the dimensions of the rectangle of maximum area that can be inscribed in the region bounded by the curve $y = 12 - x^2$ above the x-axis and the line segment from $(\sqrt{12}, 0)$ to $(-\sqrt{12}, 0)$, presuming that the sides of the rectangle are parallel to the axes. Also find the area of the largest isosceles trapezoid that can be inscribed, with one side of the trapezoid extending from $(\sqrt{12}, 0)$ to $(-\sqrt{12}, 0)$.

C35. Among all possible canvas tents of conical shape, with specified volume and no flooring material, what should be the ratio of height h to radius r to minimize the amount of material used? (The formulas for volume and lateral surface area needed here are $V = \pi r^2 h/3$ and $S = \pi r\sqrt{r^2 + h^2}$.)

C36. (An application of geometry to algebra.) Prove that

$$\sqrt{a_1{}^2 + b_1{}^2} + \sqrt{a_2{}^2 + b_2{}^2} + \sqrt{a_3{}^2 + b_3{}^2} \geq \sqrt{(a_1 + a_2 + a_3)^2 + (b_1 + b_2 + b_2)^2} \tag{1}$$

holds for any real numbers a_1, a_2, a_3, b_1, b_2, b_3, with equality iff the points (a_1, b_1), (a_2, b_2), (a_3, b_3) are collinear with the origin and lie on one side of it. Another way to state this condition for equality is that the pairs a_1, b_1 and a_2, b_2 and a_3, b_3 are proportional with no two of a_1, a_2, a_3 having opposite signs.

Suggestion: Compare the sum of the distances $OP + PQ + QR$

with OR, where O is the origin and P, Q, R are (a_1, b_1), $(a_1 + a_2, b_1 + b_2)$, $(a_1 + a_2 + a_3, b_1 + b_2 + b_3)$. (The inequality (1) can be readily extended so that the left side is the sum of terms $\sqrt{a_j^2 + b_j^2}$ for $j = 1, 2, \ldots, m$, with the appropriate change on the right side. Moreover, by taking the points in 3-space rather than in the plane, so that for example (a_1, b_1) is replaced by (a_1, b_1, c_1), the square roots on the left side have the more general form $\sqrt{a_1^2 + b_1^2 + c_1^2}$, etc. Finally, the generalization to n-dimensional space follows naturally.)

C37. Let a, b, h be positive constants, with $a \neq b$. Among all trapezoids with parallel sides of lengths a and b, and distance h between them, prove that the one with least perimeter is the isosceles trapezoid.

C38. Let a, b, c, d be positive real numbers. For what real number x is

$$\sqrt{a^2 + (b - x)^2} + \sqrt{c^2 + (d + x)^2}$$

a minimum?

C39. Find the point or points Q in the plane of a regular polygon such that the sum of the perpendicular distances from Q to the sides of the polygon is a minimum.

C40. Given a point P inside an angle QOR, how should the points H on OQ and K on OR be chosen so that H, P, K are collinear and the product $PH \cdot PK$ is a minimum. *Suggestion:* Use an auxiliary circle tangent to OQ and OR. (This is a paraphrased version of problem Al from the thirty-seventh annual William Lowell Putnam Competition, reproduced here with the permission of the Mathematical Association of America.)

C41. A point P lies 1000 meters north of a point Q. Cyclist A starts moving south from P at a speed of 6 meters per second. Starting at the same time, cyclist B travels east from Q at 8 meters per second. Find the shortest distance between the two cyclists.

C42. The *diameter* of a set of points is defined as the least upper bound of the distances between pairs of points of the set. For a polygon, the least upper bound is the maximum of the distances. It is not difficult to see, in fact, that the diameter of a polygon is the maximum of the distances between pairs of vertices. Prove that among all quadrilaterals of diameter 1, the maximum possible area is $^1/_2$. Prove also that although the square with diagonal 1 has this largest possible area, it is not the unique quadrilateral of diameter 1 and area $^1/_2$.

Notes on Chapter 3

§3.3 See Schaumberger (1977) for results on the centroid going beyond problem C3.

§3.4. For an extension of Problem 2, see Wagner (1976).

The line of argument in Problem 3 associated with Figure 3.4d was suggested by Alan Hoffer. For another approach to the question in the case where the inscribed figure does not have all its vertices on the sides of the triangle, but limited to inscribed rectangles, see the paper by M. T. Bird (1971).

Problem C26 is a paraphrase of a problem from the 1976 U.S.A. Mathematical Olympiad, included here by the courtesy of Murray S. Klamkin and Samuel L. Greitzer; see their 1976 paper listed in the References.

§3.5. The first result, on the shortest distance from one point to another via a line, is called Heron's theorem, whereas the formula for the area of a triangle in terms of its sides in §1.3 is called Hero's formula. The originator is the same, a scientist who lived in Alexandria at a time not known precisely, but not later than A.D. 300.

What happens to Heron's theorem in 3-space? That is, what is the location of P on the line CD to minimize $AP + PB$ in case the points A and B and the line CD do not lie in the same plane? This question is included in the problems at the end of Chapter 11.

§3.7. The solution of Problem 1 is adapted from Pólya (1968, pp. 205, 206).

Variations and extensions of the results of this chapter can be found in the papers by Chakerian and Lange (1971), Fisk (1977), and Lange (1976).

Among all n-gons of diameter 1, which has the largest area? (The case $n = 4$ is discussed in Problem C42.) If n is odd, the answer to this question is the regular n-gon. If $n = 6$, the question, unanswered for 20 years, was settled by R. L. Graham, who showed that there is a unique hexagon of diameter 1 having largest area, and it is *not* the regular hexagon. For details see Graham (1975).

For the method of orthogonal projections and their application to extremal problems, see the paper by G. D. Chakerian, "A Distorted View of Geometry," in Mathematical Plums (Ross Honsberger, editor), Dolciani Mathematical Expositions, No. 4, Mathematical Association of America, 1979, pp. 130–150.

CHAPTER **4**

ISOPERIMETRIC RESULTS

4.1. Some Definitions. "Isoperimetric" means constant perimeter. *An isoperimetric problem* is one that asks for the region of largest area in a given class of regions, such as triangles, of a specified perimeter. It is logically equivalent to ask for the region of least perimeter in some class of regions having a given area. *The isoperimetric problem* asks for the region of largest area among *all* possible regions having a given perimeter. In 3-space, *the* isoperimetric problem is to find the geometric figure of largest volume among *all* figures having a given surface area.

Two basic isoperimetric results are proved in this chapter; first, that among all n-gons of a given perimeter the regular n-gon has largest area; and, second, that among all simple closed curves of a given length the circle encloses the largest area. A *closed curve* is one that has no endpoints, so that starting from any point P on the curve and continuing in one direction along the curve, we reach the point P again. A closed curve is *simple* if it does not cross or touch itself at any point. Thus a circle, a triangle, a rectangle, and an ellipse are simple closed curves, whereas a figure eight is closed but not simple.

The results in this chapter are proved on the assumption that there *is* a region of largest area of a given perimeter. These same results are proved again in Chapter 12 without the assumption that a solution exists. Further remarks are made in §4.5 about the existence of a solution of a problem, and also about the uniqueness of the solution. For the present, the reader is reminded that solutions need not exist: In Problem C10 it was established that there does not exist a figure of maximum area among nonconvex quadrilaterals having a specified perimeter.

No complete analysis is given here of the length of a curve or the area enclosed by a simple closed curve. The following remarks will suffice as intuitive observations on the matter. Length or distance is first defined on a straight line, beginning with some arbitrarily chosen unit of length. The *length of a curve* from point A to point B is defined as follows. Choose points $P_1, P_2, \ldots, P_{n-1}$ on the curve between A and B, and denote A by P_0 and B by P_n. If the sum of the lengths of the straight line segments (or chords)

$$P_0P_1 + P_1P_2 + P_2P_3 + \cdots + P_{n-1}P_n \tag{1}$$

approaches a limit as n increases indefinitely in such a way that the largest term in the sum (1) tends to zero, then this limit is said to be the length of the curve AB. The limit need not exist in cases of very irregular curves, in which case length is not defined.

Area is first defined for a unit square, that is, a square whose sides are of unit length. From this it is easy to define and calculate the area of a rectangle, a parallelogram, a triangle, and so in general any polygon with n sides. The area of a plane region can then be defined as the greatest lower bound of the set S of numbers obtained as follows: each number is the sum of the areas of equal nonoverlapping squares that cover the region.

The analysis above of length and area is intuitive in nature, and lacks depth. It suffices for our purposes, however, for the following reason. As a first step in the analysis of both the n-gon of largest area and the region in the plane of largest area, of a specified perimeter in each case, we show that we may confine attention to *convex n*-gons and convex regions. A *convex region R* in the plane is one such that the line segment joining any two points of R is contained in R. The boundary of a bounded convex region is a simple closed curve which *does* have a length, and the region *does* have an area. (The word "bounded" must be inserted here so as to exclude unbounded regions like the entire plane or the first quadrant, which clearly do not have finite areas or perimeters.) Because of these results about the existence of areas and perimeters of bounded convex regions, we shall presume without repeatedly asserting it that the curves referred to in the results in §4.3 are restricted to those having lengths and areas as needed in the arguments. For proofs of these results on existence the reader is referred to the book by Russell V. Benson (1966, pp. 142, 158) in the bibliography.

A polygon is *regular* if its sides are equal and its interior angles are equal. A regular polygon is inscribable in a circle, and also circumscribable about a circle. The first statement means that there is a circle passing through all the vertices; the second means that there is a circle having every side of the polygon as a tangent line.

D1. (a) If an n-gon inscribed in a circle has equal sides, must it be regular? (b) The same question for an n-gon circumscribed about a circle.

D2. (a) If an n-gon inscribed in a circle has equal interior angles, must it be regular? (b) The same question for an n-gon circumscribed about a circle.

4.2. Polygons.

THEOREM 4.2a. *If an n-gon P is not regular, there is another n-gon P_1 with the same perimeter and larger area.*

This has already been proved in the cases $n = 3$ and $n = 4$ in the preceding chapter.

If P is not convex we look at the "convex hull" of P, defined as the smallest convex set containing P. This is illustrated in Figure 4.2a,

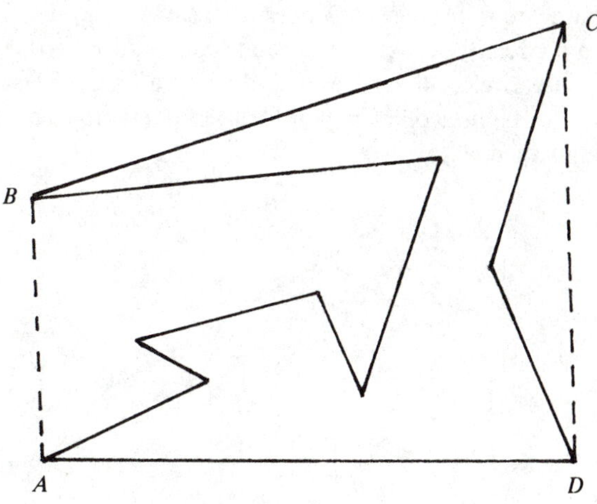

FIG. 4.2a

where P is the polygon with the solidly drawn sides, and the convex hull of P is the quadrilateral $ABCD$ and its interior. In general, the convex hull H of a nonconvex polygon P is a polygonal region with fewer vertices, shorter perimeter, but larger area than P. Convert the convex hull H into an n-gon by introducing a sufficient number of vertices along the boundary to bring the total up to n. This does not alter the perimeter or the area of H. Then define P_1 to be the n-gon that is geometrically similar to the boundary of H, having the same perimeter as P. Thus P_1 has larger area than H, and so also larger area than P.

Now we turn to the case where the given polygon P is convex. We separate the argument into two subcases: first, where not all sides of P are of equal length; second, where all sides are equal. In case not all sides of P are of equal length, there must be two adjacent unequal sides, say $A_1A_2 \neq A_2A_3$, where A_1, A_2, A_3 are consecutive vertices. By Theorem 3.2b we can locate the point K on the same side of A_1A_3 as A_2, such that $A_1K = KA_3$ and $A_1K + KA_3 = A_1A_2 + A_2A_3$, as illustrated in Figure 4.2b. Then the triangle A_1KA_3 has larger area than $A_1A_2A_3$. Therefore the polygon P_1 obtained from P by replacing the vertex A_2 by K satisfies the conditions of the theorem.

Finally, suppose that the n-gon P is convex and all sides are equal in length. Since P is not regular, the vertices do not lie on a circle. Hence there are four consecutive vertices, say A_1, A_2, A_3, A_4, that do not lie on a circle, because if every four consecutive vertices were inscribable in a circle, all vertices would lie on a circle. The reason for this is that three noncollinear points determine a unique circle passing through them.

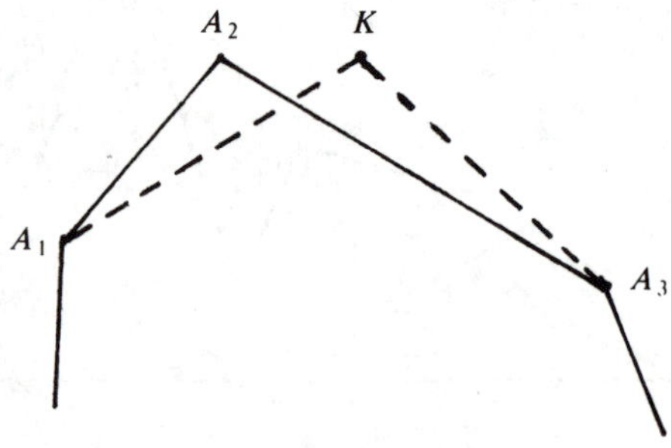

FIG. 4.2b

Now by Theorem 3.3b the quadrilateral inscribable in a circle, with sides of the same lengths as quadrilateral $A_1A_2A_3A_4$, has larger area. Such an inscribable quadrilateral can be taken as $A_1A_2'A_3'A_4$ with the vertices A_2 and A_3 replaced by A_2' and A_3' as shown in Figure 4.2c. All the sides here are equal:

$$A_1A_2 = A_2A_3 = A_3A_4 = A_1A_2' = A_2'A_3' = A_3'A_4.$$

The desired n-gon P_1 is simply P with the vertices A_2 and A_3 replaced by A_2' and A_3', respectively.

THEOREM 4.2b. *Consider the set of all n-gons with a specified perimeter c. Assuming that there is one of maximum area, then there is only one, and it is the regular polygon with n sides.*

Let P be an n-gon with perimeter c having maximum area. By Theorem 4.2a the n-gon P must be regular, since otherwise there is an n-gon P_1 with larger area and perimeter c. Hence P is unique.

4.3. The Isoperimetric Theorem. The problem of enclosing the largest area with a curve of given length can be traced through Roman legend to the so-called problem of Dido. Shipwrecked, Queen Dido is reputed to have asked the local residents for a parcel of land along the seashore for herself and her retinue. She was granted as much land as could be encompassed by the hide of an ox. Cleverly

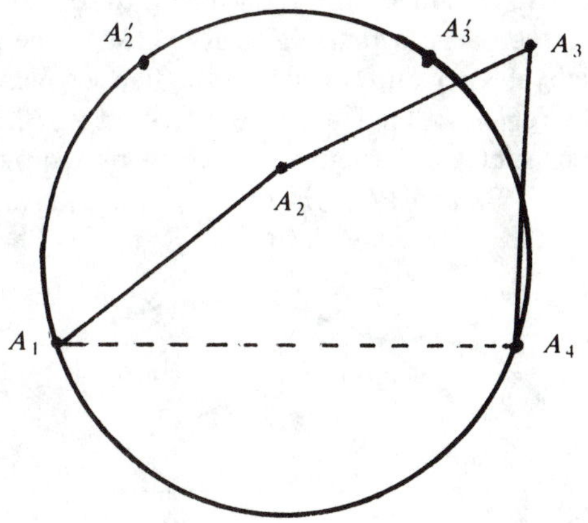

FIG. 4.2c

cutting the oxhide into very narrow strips, she joined them to make one very long strip and marked off quite a large region, thus founding the city of Carthage. The problem is: What curve should Dido follow with her length of oxhide, from one point on the seashore to another, so as to demarcate the largest possible area? (The shore line is assumed to be straight.) The answer is given in the following result.

THEOREM 4.3a (The Dido Problem). *For any positive constant c consider the problem of finding the curve of length c that encloses the largest possible area between itself and a given straight line. (This is illustrated in Figure 4.3a where the curve of length c reaches from A to B enclosing an area against the straight line.) If the curve is not a semicircle, then it can be replaced by another curve of length c enclosing a larger area. Hence if there is a curve enclosing the largest area, it must be a semicircle.*

This is similar to Problems C8 and C9 about the farmer enclosing a maximum area by fencing three sides of a rectangular region, or quadrilateral region, along a straight river bank. The difference now is that the farmer has flexible fencing that can be shaped into any desired curve.

If the curve from A to B, as in Figure 4.3a, is not a semicircle, then there is a point P on the curve at which the angle $APB \neq 90°$. Then joining P to A and B, the total region under consideration is divided into three parts: a first region R_1 between the line segment AP and the curve; a region R_2 between the line segment BP and the curve; and the region R_3 enclosed by the triangle APB.

Now we construct another curve, obtained by redrawing the arcs AP and PB as $A'P'$ and $P'B'$ in Figure 4.3b, with angle $A'P'B'$

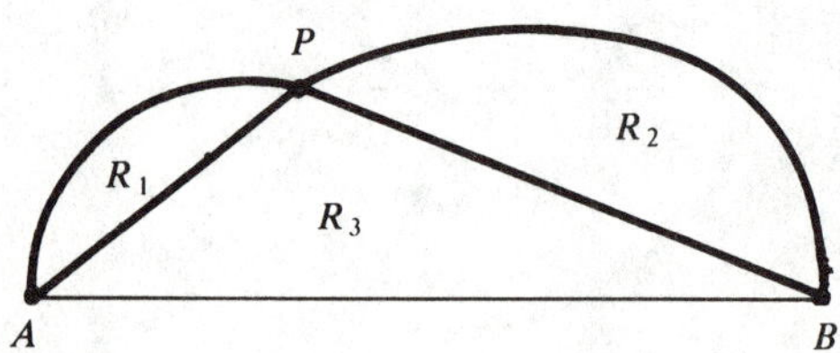

FIG. 4.3a

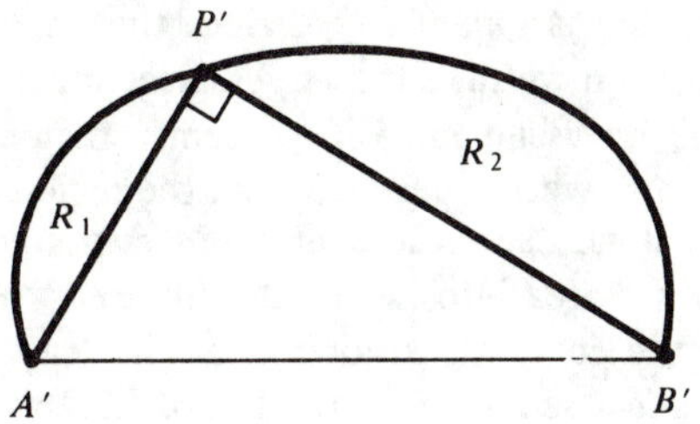

FIG. 4.3b

$= 90°$. To spell this out in more detail, the line segments $A'P'$ and $P'B'$ are drawn at right angles, with $A'P' = AP$ and $P'B' = PB$. On $A'P'$ we construct a region congruent to R_1, and on $P'B'$ a region congruent to R_2. The area of triangle $A'P'B'$ is larger than that of APB by Theorem 3.2c. Hence the curve from A' to P' and on to B' encloses a larger area against the straight line, but has the same length as the curve APB in Figure 4.3a.

The main result now follows by an easy application of the reflection principle.

THEOREM 4.3b (The Isoperimetric Theorem). *Given any simple closed curve C_1, not a circle, there is another curve C_2 of the same length that encloses a larger area. Hence if there is a simple closed curve of a specified length that encloses a maximum area, it must be a circle.*

If the region R_1 bounded by C_1 is not convex, we argue as in the proof of Theorem 4.2a. As suggested by Figure 4.2a, we pass to the convex hull of R_1, that is, the smallest convex set containing R_1. If C is the boundary of this convex hull, then the length of C is less than the length of C_1, but C encloses a larger area. Then we define C_2 as the curve that is geometrically similar to C, having the same length as C_1. This completes the proof of the theorem in this case, and we now turn to the other possibility, namely, that the given curve C_1 encloses a region that is convex.

For convenience let k denote the length of the curve C_1. Take any point P on C_1, and let Q be the point half-way around the perimeter, as shown in Figure 4.3c. Thus the arc PQ, measured in either direction from P to Q along the curve, has length $k/2$. Now suppose first that the straight line segment PQ divides the entire area into two unequal pieces. Then we take the larger piece together with its mirror image in the line segment PQ to form a region bounded by a curve C_2 of length k, with larger area than the region bounded by C_1.

On the other hand, suppose that the straight line segment PQ separates the entire area into two equal sub-areas, as shown in Figure 4.3c. Since the curve C_1 is not a circle, at least one of the arcs from P to Q is not a semicircle, say the arc PSQ as in Figure 4.3c. Then, using Theorem 4.3a, we can replace the arc PSQ by one of the same length that encloses a larger area against a straight line, and this with its mirror image about the straight line gives the simple closed curve C_2 that we want. That is, C_2 has length k by the construction, but it encloses a larger area than C_1.

The ingenious proofs of Theorems 4.3a and b were given in 1836 by Jakob Steiner, a professor at the University of Berlin who greatly enriched the field of geometry with his remarkably clever methods. During that period, however, more rigorous standards of proof were being developed in mathematics. Although Steiner thought he had proved the isoperimetric theorem, Dirichlet pointed out that he had not proved the actual existence of a maximum, only that, if a curve exists enclosing a maximum area, it must be the circle. Much work was done subsequently to fill the gap in the argument. We give one method of doing this in Chapter 12, using a method that supplements, but does not replace, the work of Steiner.

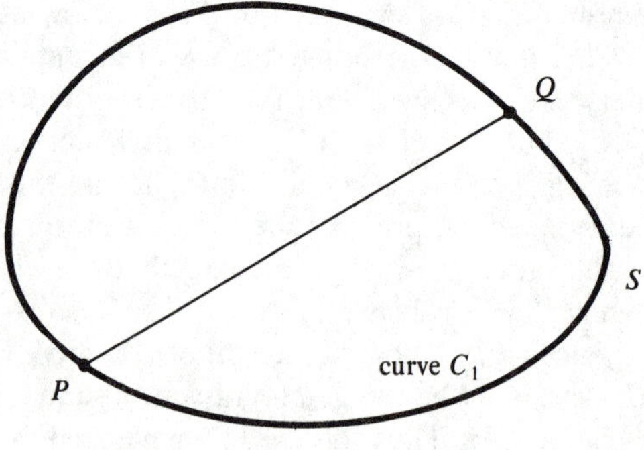

FIG. 4.3c

D3. Given a line segment AB and a constant c exceeding the length of AB, prove that among all arcs from A to B of length c, the circular arc encloses with the segment AB the maximum area. *Suggestion:* Draw not only the circular arc from A to B of length c but also the rest of the circle on the other side of AB.

4.4. The Isoperimetric Quotient. The isoperimetric theorem can be stated in the following way:

For every region in the plane with area A and perimeter L the inequality

$$4\pi A \leq L^2 \quad or \quad 4\pi A/L^2 \leq 1 \tag{1}$$

holds, with equality iff the region is a disk, that is, a circle and its interior.

The reason for this is easy to see. The circle with perimeter L has radius $L/2\pi$, and so has area $\pi(L/2\pi)^2 = L^2/4\pi$. Since the circle has largest area among curves of a given perimeter, the area of any other such curve must be less than $L^2/4\pi$, and this is what the inequality (1) states

The ratio $4\pi A/L^2$ has been called the *isoperimetric quotient* (or IQ) of a region in the plane. The factor 4π is said to "normalize" the ratio, so that a circle has IQ equal to 1, and all other IQ's are smaller, by (1) above. Furthermore the quotient as defined has the property that *similar figures have the same IQ's.* Take two similar triangles, for example, one with sides of lengths a, b, c and the other with sides a_1, b_1, c_1. If r is the ratio of corresponding lengths, then

$$a = ra_1, \quad b = rb_1, \quad c = rc_1 \quad \text{and} \quad L = rL_1,$$

where L and L_1 are the perimeters of the two triangles. If the areas of the triangles are A and A_1, respectively, then $A = r^2A_1$, as can be easily seen by using any of the standard formulas for the area of a triangle. From $L = rL_1$ and $A = r^2A_1$ we get $A/L^2 = A_1/L_1{}^2$, and hence the IQ's are equal.

The same result holds for any two similar figures, because the basic property of such figures is that any two corresponding lengths have the same ratio. Thus $L = rL_1$ holds just as in the case of the triangle, and corresponding areas A and A_1 again satisfy $A = r^2 A_1$, because area is a two-dimensional concept, always calculated from products of pairs of lengths.

For example, all equilateral triangles have the same IQ, $\pi\sqrt{3}/9$ or .605 to 3 decimal places. Since the IQ is a measure of how large an area is enclosed by a given perimeter length, with 1 as the largest possible value, it is interesting to compare different shapes. Here is a brief list of standard figures and their IQ's to 3 decimal places:

Circle	1.0
Regular octagon	.948
Regular hexagon	.907
Regular pentagon	.865
Square	.785
Sector, angle 2 radians	.785
Sector, angle 90°	.774
Rectangle, 3 by 2	.754
Semicircle	.747
Rectangle, 2 by 1	.698
Equilateral triangle	.605
Rectangle, 3 by 1	.589
Isosceles right triangle	.539
Triangle, sides 1, 2, $\sqrt{3}$	.486

By "sector" is meant a sector of a circle having the angle indicated at the center. The sector with angle 90° is also called a quadrant. The sector with angle 2 radians has a special property, featured in one of the problems at the end of this section. This sector has the same IQ as the square, as is seen in the table; the exact value is $\pi/4$ in both cases.

The table includes the regular polygons with 3, 4, 5, 6, and 8 sides. It is to be expected that the regular polygon with $n + 1$ sides has a higher IQ than the regular n-gon. This is so, and the techniques needed to establish this result are developed in the next chapter.

D4. Consider a sector of a circle with an angle θ between the two radii. What should be the size of θ to maximize the isoperimetric quotient? Suggestion: If θ is in radians, and r is the radius, the perimeter is $2r + r\theta$ and the area is $r^2\theta/2$.

D5. Verify that a square and a sector with angle 2 radians have the same IQ.

D6. (i) Is there any parallelogram $ABCD$ having the same IQ as its triangle ABC? If so, characterize all such parallelograms. (ii) Is there any rectangle with this property?

4.5. Existence and Uniqueness. Several of the problems discussed in this book can be handled more easily if the existence of a solution, or various elements in a solution, are assumed. To illustrate this point, we now give a very simple proof of the inequality of the arithmetic-geometric means of Chapter 2.

Given a positive constant c and an integer $n > 1$, there are many sets of positive numbers $x_1, x_2, \ldots, x_n$ having sum c. We assume that among such sets there is a set $a_1, a_2, \ldots, a_n$ having the largest product. It is easy to prove that these numbers are all equal. For suppose on the contrary that there are two unequal quantities, say $a_i \neq a_j$. If we replace each of a_i and a_j by $(a_i + a_j)/2$ and leave the other a's unchanged, we have a new set with the same sum c, but a larger product since $(a_i + a_j)^2/4 > a_i a_j$. This constitutes a contradiction and we conclude that $a_1 = a_a = \cdots = a_n$. The arithmetic mean of these numbers, c/n, equals their geometric mean. Any *other* set of positive numbers $x_1, x_2, \ldots, x_n$ having sum c has a smaller product and hence a smaller geometric mean. It follows that

$$(x_1 + x_2 + \cdots + x_n)/n > (x_1 x_2 \cdots x_n)^{1/n}$$

for all sets of positive numbers with sum c, except the special set with all the numbers equal.

(The proof above is taken in paraphrased form from an article on extremal questions. The writer of the article used the existence assumption "... and let $a_1, a_2, \ldots, a_n$ be the set having the greatest product" without identifying it as a powerful hypothesis. The same proof is given in Courant and Robbins (1947, p. 364) but with the assumption of existence clearly labeled and analyzed.)

As another illustration of an easier solution assuming existence, consider Theorem 3.3b, that among all quadrilaterals of a given perimeter the square has the largest area. If we assume the existence of a quadrilateral of largest area, any two adjacent sides must be equal by Theorem 3.2b. Hence all four sides are equal, and it is an easy argument to show that each interior angle must be 90° for largest area. Thus the assumed quadrilateral is a square.

Of course the treatment of a problem is more complete and accurate from a logical standpoint if the existence of a solution is *not* assumed. Consider the following problem, for example: Given any triangle ABC, find the inscribed triangle PQR of least perimeter. Such a minimum triangle, illustrated in Figure 4.5a, must have a simple property suggested by the following argument. Regarding Q and R as fixed points for a moment, the point P must be located on BC so that $RP + QP$ is a minimum. This question was discussed as the first example on the reflection principle in §3.5, where it was shown that in effect P must be located so that $\angle RPB = \angle QPC$. Similarly it must be that

$$\angle QRA = \angle PRB \quad \text{and} \quad \angle RQA = \angle PQC.$$

Thus we conclude that the inscribed triangle PQR of least perimeter is the one having the property that its sides make equal angles with the sides of triangle ABC in the sense just indicated.

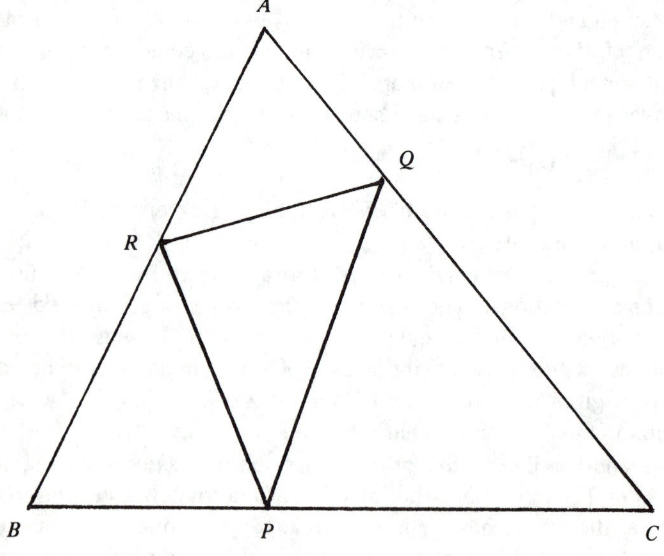

FIG. 4.5a

However, as a solution to the problem this has some gaps. Only in an acute-angled triangle *ABC* does the triangle *PQR* exist having the equal angle property described above. (This problem is discussed more fully in Chapter 9, where a different solution is given, a solution that enables us to construct the triangle *PQR* when it exists.) Furthermore, assuming the existence of a triangle *PQR* with the equal angle properties, how do we know that there is only one? The language, "We conclude that the inscribed triangle *PQR* of least perimeter is *the* one," begs the question not only of the existence of such a triangle but also of the uniqueness. The triangle *PQR*, if it exists, *is* unique, but the argument above is not adequate to establish this result.

In many situations it is easy to argue that, given a configuration of a certain type, a larger one can be constructed. (This argument was used repeatedly in §§4.2 and 4.3) If the larger one can be constructed in all cases except one, does it follow that this exceptional one is the largest? The following example reveals that the answer is no. Consider the set of positive integers or natural numbers, 1, 2, 3, 4, ... Corresponding to each member a of the set, there is a larger one a^2, with one exception. The exception, of course, occurs in case $a = 1$, in which case a^2 is not larger. But it is hardly the case that 1 is the largest positive integer. There is no largest.

Problem C10 has been cited as an illustration of a problem having no solution. A famous example of this phenomenon is the *Kakeya problem:* Of all plane regions R with the property that one can turn a unit segment through a full 360° in R, which has the smallest area? A disk, that is, a circle together with its interior, of diameter one is an example of such a region. It was shown by A. S. Besicovitch that there is *no* such region of least area: Given any positive number β, no matter how small, there is a region with the prescribed property that has an area $< \beta$.

In more advanced books on mathematics the question of existence of a solution is argued in many cases by the "compactness" of the set involved. For example, the n-gons in §4.2 can be described in terms of the coordinates of their vertices. Since each vertex requires two coordinates x and y in the plane, we have $2n$ coordinates in all, and these may be restricted to a compact set of "points" in $2n$-dimensional space. This assures us of the existence of an n-gon of maximum area among those of a specified perimeter. This is a consequence of the general result that a continuous real-valued function f defined on a nonempty compact subset of R^n is bounded and attains a maximum value and a minimum value; that is, the function attains its least upper bound and its greatest lower bound. The space R^n is the direct product space $R \times R \times \cdots \times R$ of the real line R with itself n times. Since any closed, bounded set in R^n is compact, this result can also be stated in terms of continuous real-valued functions on closed, bounded sets. Arguments of this sort based on closed sets or compact sets, although applicable to the question of the n-gon of maximum area among those having a specified perimeter, cannot readily be used to assure the existence of a simple closed curve surrounding the largest area among curves of a specified perimeter. In any case, all these considerations are beyond the scope of this book. Our remarks are intended to draw attention to more sophisticated ways of deciding the question of existence for the benefit of readers who may want to pursue such matters. Readings on compact sets are given in the notes at the end of the chapter.

Finally, we turn to the less sophisticated question of the uniqueness of a solution. If there is a solution, might there be more than one? The answer is yes, as we would expect. One easy example, solved in Chapter 7, is the problem of finding the quadrilateral of largest area inscribed in an ellipse. *The* quadrilateral? There are infinitely many of them, it turns out, even infinitely many incongruent ones. One of these quadrilaterals of maximum area is the rectangle asked for in a standard problem in calculus books, the rectangle of maximum area inscribed in an ellipse with sides parallel to the axes of the ellipse.

The uniqueness question, then, is simply a matter of determining whether a problem has only one solution, or more than one.

Notes on Chapter 4

§4.1. More elaborate discussions of lengths, area, and convexity can be found in Benson (1966), Chakerian (1969), Coolidge (1953), Klee (1949 and 1971), Rado (1943), Valentine (1964), Whyburn (1942), and Yaglom and Boltyanskii (1961).

§4.4. Our definition of isoperimetric quotient follows that of Pólya (1954, p. 179). Fejes Tóth (1964, p. 157) uses the ratio L^2/A, and the ratio S^3/V^2

for the isoperimetric quotient of a solid figure with volume V and surface area S. The Pólya version of this would be $36\pi V^2/S^3$, with value 1 for a sphere and smaller values for other solid figures. For a proof of this property of a sphere, see Benson (1966, Chapter 4).

§4.5. The existence of a polygon of largest area among n-gons of a given perimeter can be demonstrated by an argument using compactness, for example, as given by Courant and Robbins (1947, p. 375) and by Fejes Tóth (1972, p. 5). For explanations of compactness and its use in establishing the existence of maxima and minima, see Boas (1960, pp. 38–44), Hewitt (1960), and Rudin (1976, p. 89). Compactness is an analytic concept not included in this book. In Chapter 12 the existence of a unique n-gon of largest area is demonstrated by an elementary method—elementary, but not simple.

For other approaches to the problems of this chapter, see DeMar (1975) and Kazarinoff (1964).

The following problem has attracted wide attention: What is the smallest possible area of a *convex* region R in the plane with the property that a segment of length 1 can be rotated through 360° without moving outside R? The circle of diameter 1 is a region having the prescribed property, but it is not of least area. The least area occurs in the case of an equilateral triangle of altitude 1, with area $\sqrt{3}/3$; for a proof see Yaglom and Boltyanskii (1961, pp. 60, 221, 222).

The same question, without the requirement of convexity, is called the Kakeya problem. The surprising solution by A. S. Besicovitch is that there is no least area. That is, there is a nonconvex region of arbitrarily small area within which a segment of length 1 can be rotated through 360°. For a proof, see Yaglom and Boltyanskii (1961, pp. 226, 227).

CHAPTER **5**

BASIC TRIGONOMETRIC INEQUALITIES

5.1. A New Direction. In this and the next two chapters we take a different direction. Our principal goal is to determine the polygons of largest area and perimeter that can be inscribed in, and circumscribed about, a circle. This is the topic of Chapter 6, and in Chapter 7 we look at similar questions for the ellipse. One of the simplest approaches to these problems is to use trigonometric sums, which arise quite naturally from the angles subtended at the center of a circle by any inscribed or circumscribed polygon.

The trigonometric inequalities we develop are special cases of results by Jensen, which are given in §5.3 and then applied broadly in §5.4. The problems for the reader at the end of §5.4 illustrate the power of the method. Questions of the maximum and minimum values of various sums and products of the trigonometric functions of the angles of a triangle are usually asked one at a time in quite isolated fashion in the problem sections of the popular journals. The Jensen inequalities provide a good example of the unification of mathematics that can be achieved by a general theoretical method.

It is of interest to note that J. L. W. V. Jensen (1854–1925), after whom the central results of this chapter are named, was the managing chief engineer of the telephone network of Denmark. A member of the Academy of Science, Jensen was a pioneer in the study of convex functions.

5.2. Some Trigonometric Inequalities. The arguments in subsequent sections are based on some simple results about the trigonometric functions.

THEOREM 5.2a. *If α and β are angles satisfying the conditions $0 \le \alpha \le 180°$ and $0 \le \beta \le 180°$, then*

$$\sin\alpha + \sin\beta \le 2\sin\frac{\alpha+\beta}{2}$$

with equality iff $\alpha = \beta$.

To prove this we use the identity

$$\sin\alpha + \sin\beta = 2\sin\frac{\alpha+\beta}{2}\cos\frac{\alpha-\beta}{2}.$$

Now $\alpha - \beta$ satisfies $-180° \le \alpha - \beta \le 180°$, and it follows that $\cos[(\alpha - \beta)/2]$ is positive or zero, with maximum value 1 iff $\alpha = \beta$. This proves the theorem.

Next we generalize this result to n angles, as follows.

THEOREM 5.2b. *If $\alpha_1, a_2, \ldots, \alpha_n$ are angles satisfying the conditions $0 \le \alpha_j \le 180°$ for all $j = 1, 2, \ldots, n$ then*

$$\sin\alpha_1 + \sin\alpha_2 + \cdots + \sin\alpha_n \le n\sin\frac{\alpha_1+\alpha_2+\cdots+\alpha_n}{n}, \tag{1}$$

with equality iff $\alpha_1 = \alpha_2 = \cdots = \alpha_n$.

It is clear that (1) becomes an equality in case the α_j are all equal; so we assume that the angles are not all equal and prove (1) with a strict inequality. Our argument is by a special induction on n, in a fashion similar to the Cauchy proof of the inequality of the arithmetic-geometric means in §2.5.

Let P_n denote the proposition that if $\alpha_1, \alpha_2, \ldots, \alpha_n$ are angles, not all equal, satisfying the conditions of the theorem, then (1) holds with strict inequality. By Theorem 5.2a above we know that P_2 holds. To prove that P_n holds for $n \ge 3$ we establish two results.

FIRST RESULT. *If the proposition P_n holds for any integer $n \ge 3$, then P_{n-1} also holds.*

SECOND RESULT. *If the proposition P_n holds for any integer $n \ge 2$, then P_{2n} holds also.*

There is no need for any argument that these two results establish P_n for all integers $n \geq 2$, because this would be a repetition of the logic as set forth in the proof in Section 2.5.

Proof that P_n implies P_{n-1}. We assume the proposition P_n, and we want to prove P_{n-1}. Let $\alpha_1, \alpha_2, \ldots, \alpha_{n-1}$ be $n-1$ angles not all equal and define

$$\alpha_n = (\alpha_1 + \alpha_2 + \cdots + \alpha_{n-1})/(n-1). \tag{2}$$

Note that

$$\alpha_1 + \alpha_2 + \cdots + \alpha_{n-1} + \alpha_n = n(\alpha_1 + \alpha_2 + \cdots + \alpha_{n-1})/(n-1)$$

and

$$(\alpha_1 + \alpha_2 + \cdots + \alpha_{n-1} + \alpha_n)/n$$
$$= (\alpha_1 + \alpha_2 + \cdots + \alpha_{n-1})/(n-1) = \alpha_n.$$

Hence P_n, applied to this set of n angles, gives

$$\sin \alpha_1 + \sin \alpha_2 + \cdots + \sin \alpha_{n-1} + \sin \alpha_n < n \sin \alpha_n.$$

Subtracting $\sin \alpha_n$ we get

$$\sin \alpha_1 + \sin \alpha_2 + \cdots + \sin \alpha_{n-1} < (n-1) \sin \alpha_n,$$

and this is P_{n-1} because of the definition (2).

Proof that P_n implies P_{2n}. Assuming the proposition P_n we want to prove the proposition P_{2n}. That is, we want to prove that if $\alpha_1, \alpha_2, \ldots, \alpha_{2n}$ are any angles, not all equal, satisfying $0 \leq \alpha_j \leq 180°$ then

$$\sin \alpha_1 + \sin \alpha_2 + \cdots + \sin \alpha_{2n}$$
$$< 2n \sin \frac{\alpha_1 + \alpha_2 + \cdots + \alpha_{2n}}{2n}. \tag{3}$$

By rearranging notation if necessary we may assume that $\alpha_1 \neq \alpha_2$. Applying P_n first to the set $\alpha_1, \alpha_2, \ldots, \alpha_n$ and then to the set $\alpha_{n+1}, \alpha_{n+2}, \ldots, \alpha_{2n}$, we get

$$\sin\alpha_1 + \sin\alpha_2 + \cdots \sin\alpha_n < n\sin\frac{\alpha_1+\alpha_2+\cdots+\alpha_n}{n} \tag{4}$$

and

$$\sin\alpha_{n+1} + \sin\alpha_{n+2} + \cdots + \sin\alpha_{2n} \le n\sin\frac{\alpha_{n+1}+\alpha_{n+2}+\cdots+\alpha_{2n}}{n}. \tag{5}$$

This last inequality is not strict because it is possible that $\alpha_{n+1} = \alpha_{n+2} = \cdots = \alpha_{2n}$. Define α and β by

$$\alpha = (\alpha_1 + \alpha_2 + \cdots + \alpha_n)/n,$$

$$\beta = (\alpha_{n+1} + \alpha_{n+2} + \cdots + \alpha_{2n})/n,$$

so that $0 \le \alpha \le 180°$, $0 \le \beta \le 180°$, and $(\alpha+\beta)/2 = (\alpha_1 + \alpha_2 + \cdots + \alpha_{2n})/2n$. Then the addition of (4) and (5) gives the strict inequality

$$\sin\alpha_1 + \sin\alpha_2 + \cdots + \sin\alpha_{2n} < n(\sin\alpha + \sin\beta).$$

But

$$\sin\alpha + \sin\beta \le 2\sin\frac{\alpha+\beta}{2}$$

and (3) follows from this.

Example. If α, β, γ are the angles of any triangle, what are the maximum and minimum values of $\sin\alpha + \sin\beta + \sin\gamma$?

A direct application of Theorem 5.2b gives

$$\sin\alpha + \sin\beta + \sin\gamma \le 3\sin\frac{\alpha+\beta+\gamma}{3} = 3\sin 60° = 3\sqrt{3}/2. \tag{6}$$

So the maximum is $3\sqrt{3}/2$, occurring iff the triangle is equilateral.

On the other hand, it is not difficult to see that there is no minimum, as follows. For any triangle, $\sin \alpha$, $\sin \beta$, and $\sin \gamma$ are positive. But the sum is as small as we please by taking α and β near zero, and γ near 180° satisfying $\gamma = 180° - \alpha - \beta$. Hence 0 is the greatest lower bound, but not the minimum of $\sin \alpha + \sin \beta + \sin \gamma$.

Some further consequences can be obtained easily from (6) above. The formula $A = \frac{1}{2}ab \sin \gamma$ and its analogs for the area A of a triangle can be written as

$$\sin \gamma = 2A/ab, \quad \sin \beta = 2A/ac, \quad \sin \alpha = 2A/bc.$$

Substituting in (6), we get, after a simple algebraic rearrangement,

$$A \leq \frac{3\sqrt{3}\, abc}{4(a+b+c)}.$$

But Hero's formula gives A in terms of the sides, so by squaring we get

$$s(s-a)(s-b)(s-c) \leq 27(abc)^2/[16(a+b+c)^2].$$

We multiply by $16(a+b+c)^2$ and write s in the form $(a+b+c)/2$ to get

$$(a+b+c)^3(b+c-a)(c+a-b)(a+b-c) \leq 27(abc)^2.$$

But $27abc \leq (a+b+c)^3$ by the AM-GM inequality, so we can replace $(a+b+c)^3$ by the possibly smaller value $27abc$ to get

$$27abc(b+c-a)(c+a-b)(a+b-c) \leq 27(abc)^2.$$

Cancellation of $27abc$ yields the inequality

$$(b+c-a)(c+a-b)(a+b-c) \leq abc. \qquad (7)$$

This inequality has been proved in case a, b, c are the lengths of the sides of a triangle. However, in case a, b, c are positive numbers not the sides of any triangle, such as $a = 2$, $b = 5$, $c = 8$, the inequality (7) holds because the left side is negative or zero. Thus (7) holds for all positive numbers a, b, c with equality iff $a = b = c$.

It will be helpful to have analogs of Theorems 4.2a and 4.2b for the other trigonometric functions. But there is no need to work out all the details case by case, because the idea in the proof of Theorem

5.2b can be put in a general form out of which the special cases emanate. This generalization is the subject of the next section.

5.3. The Jensen Inequalities. The trigonometric result in Theorem 5.2b can be put in a general setting as follows.

THEOREM 5.3a (Jensen). *Suppose that a function $f(x)$ has the property*

$$f(\alpha)+f(\beta)\le 2f\left(\frac{\alpha+\beta}{2}\right) \tag{1}$$

for every pair of numbers in the domain of the function f, where the domain is an interval on the real number line. Suppose that equality holds only in case $\alpha=\beta$. Then for any numbers $\alpha_1, \ldots, \alpha_n$ in the domain

$$f(\alpha_1)+f(\alpha_2)+\cdots+f(\alpha_n)\le nf\left(\frac{\alpha_1+\alpha_2+\cdots+\alpha_n}{n}\right), \tag{2}$$

with equality iff the α's are all equal. Furthermore, an analogous result holds with the inequalities reversed in (1) *and* (2).

The proof of this result is exactly the same as the proof of Theorem 5.2b, with the sine function replaced everywhere by the function f. The last part of Theorem 5.3a, with the inequalities reversed, is proved in the same way again, so the details are omitted.

A *convex function* is one satisfying the inequality $f(\alpha)+f(\beta)\ge 2f((\alpha+\beta)/2)$ for all α, β in the domain of the function. Because of this definition, Theorem 5.3a in a more general form is called Jensen's inequality on convex functions. It is convenient to have also a formulation of the result as a product, as follows.

THEOREM 5.3b. *Suppose that a function f has the property*

$$f(\alpha)\cdot f(\beta)\ge\left[f\left(\frac{\alpha+\beta}{2}\right)\right]^2 \tag{3}$$

for all α and β in the domain of f, with equality only in case $\alpha=\beta$.

Assume that the domain of f is an interval on the real number line, and that the range is a set of positive numbers. Then

$$f(\alpha_1)\cdot f(\alpha_2)\cdots f(\alpha_n) \geq \left[f\left(\frac{\alpha_1 + \alpha_2 + \cdots + \alpha_n}{n}\right)\right]^n \tag{4}$$

with equality iff all the α's are equal. Furthermore, an analogous result holds if the inequalities in (3) *and* (4) *are reversed.*

Our use for these results is principally in the case $n = 3$; so we restrict attention to the proof for that value of n and also $n = 4$. Let $\alpha, \beta, \gamma, \delta$ be four numbers in the domain of f. If they are all equal

$$f(\alpha)f(\beta)f(\gamma)f(\delta) \geq \left[f\left(\frac{\alpha + \beta + \gamma + \delta}{4}\right)\right]^4 \tag{5}$$

holds with equality. Now suppose that $\alpha, \beta, \gamma, \delta$ are not all equal, say $\alpha \neq \beta$. Then (3) holds with a strict inequality, and if we multiply it by the known result

$$f(\gamma)f(\delta) \geq \left[f\left(\frac{\gamma + \delta}{2}\right)\right]^2, \tag{6}$$

we get

$$f(\alpha)f(\beta)f(\gamma)f(\delta) > \left[f\left(\frac{\alpha + \beta}{2}\right)\cdot f\left(\frac{\gamma + \delta}{2}\right)\right]^2.$$

Now we can apply (3) to the product in square brackets here, and so get (5) with inequality, thus completing the proof in case $n = 4$.

For the case $n = 3$, we start with α, β, γ in the domain of f. Again setting aside the case where $\alpha = \beta = \gamma$, we want to prove

$$f(\alpha)f(\beta)f(\gamma) > \left[f\left(\frac{\alpha + \beta + \gamma}{3}\right)\right]^3 \tag{7}$$

in case α, β, γ are not all equal. Define δ as the arithmetic mean of α, β, γ thus $\delta = (\alpha + \beta + \gamma)/3$, and observe that $\alpha, \beta, \gamma, \delta$ are not all equal. Then we have

$$(\alpha + \beta + \gamma + \delta)/4 = [\alpha + \beta + \gamma + (\alpha + \beta + \gamma)/3]/4$$

$$= (\alpha + \beta + \gamma)/3.$$

Now use (5) as a strict inequality, with δ defined as above. This gives

$$f(\alpha)f(\beta)f(\gamma)f\left(\frac{\alpha + \beta + \gamma}{3}\right) > \left[f\left(\frac{\alpha + \beta + \gamma}{3}\right)\right]^4.$$

By assumption, $f((\alpha + \beta + \gamma)/3)$ is positive, so it can be divided out to give the desired result.

Second proof. Another way to prove Theorem 5.3b is based on logarithms, for those readers who are familiar with this topic. Taking logarithms in (3) we get

$$\log f(\alpha) + \log f(\beta) \geq 2 \log f\left(\frac{\alpha + \beta}{2}\right).$$

Then by Theorem 5.3a applied to the function $\log f$ we conclude that

$$\log f(\alpha_1) + \log f(\alpha_2) + \cdots + \log f(\alpha_n)$$
$$\geq n \log f\left(\frac{\alpha_1 + \alpha_2 + \cdots + \alpha_n}{n}\right).$$

But this is just the logarithmic form of the conclusion in Theorem 5.3b.

Among the special cases of Theorem 5.3b, the application to the trigonometric functions are of considerable interest. For example, the inequality

$$\sin \alpha \sin \beta \leq \sin^2 \frac{\alpha + \beta}{2} \tag{8}$$

holds for angles α and β between 0 and 180°, with equality here iff $\alpha = \beta$. (The proof of this is left to the reader as Problem E1 at the end of the next section.) Formula (8) is just the hypothesis (3) in Theorem 5.3b with the inequality reversed, so we can conclude that

$$\sin\alpha\sin\beta\sin\gamma \le \sin^3\frac{\alpha+\beta+\gamma}{3} \tag{9}$$

for angles between 0 and 180° with equality iff $\alpha = \beta = \gamma$. Similar conclusions can also be drawn for products of more than three terms of the type $\sin\alpha$.

Inequality (9) also holds for cosines, at least for acute angles α, β, and γ. This is another consequence of Problem El in the next section.

In the case of the tangent function, it turns out that for acute angles α and β the inequality

$$\tan\alpha\tan\beta < \tan^2\frac{\alpha+\beta}{2} \tag{10}$$

holds in case $\alpha + \beta < 90°$, but (10) is reversed if $\alpha + \beta > 90°$, with equality in (10) if $\alpha + \beta = 90°$. The proofs are left to the reader in Problem E6, with some consequences for the angles of a triangle in E7.

5.4. Other Trigonometric Functions. In Section 5.2 we proved basic inequalities for the sine function. In order to establish analogs for the other trigonometric functions, we proceed in a similar way to obtain first the results in case $n = 2$. The Jensen theorems can then be applied to get the general inequalities.

We know that $\sin\alpha + \sin\beta \le 2\sin[(\alpha+\beta)/2]$. This is extended by the use of the inequality of the arithmetic-harmonic means of Problem B4,

$$a + b \ge \frac{4ab}{a+b},$$

with equality iff $a = b$. Replacing a and b by $\sin\alpha$ and $\sin\beta$, we see that for angles α and β satisfying $0 < \alpha < 180°$ and $0 < \beta < 180°$,

$$2\sin\frac{\alpha+\beta}{2} \ge \sin\alpha + \sin\beta \ge \frac{4\sin\alpha\sin\beta}{\sin\alpha + \sin\beta}. \tag{1}$$

Ignore the middle expression $\sin\alpha + \sin\beta$ here, and take reciprocals of the other two. Multiplying by 4, we have

$$\operatorname{cosec} \alpha + \operatorname{cosec} \beta \geq 2\operatorname{cosec} \frac{\alpha + \beta}{2}, \tag{2}$$

with equality iff $\alpha = \beta$. This holds for $0 < \alpha < 180°$ and $0 < \beta < 180°$. (Recall that since $\sin 0 = \sin 180° = 0$, cosec 0 and cosec 180° are not defined.)

Next, the inequality

$$\cos \alpha + \cos \beta \leq 2 \cos \frac{\alpha + \beta}{2} \tag{3}$$

holds for $-90° \leq \alpha \leq 90°$, $-90° \leq \beta \leq 90°$, with equality iff $\alpha = \beta$. This follows directly from Theorem 5.2a by replacing α and β by $90° - \alpha$ and $90° - \beta$, and using the well-known identity $\sin(90° - \alpha) = \cos \alpha$. Note that the condition $0 \leq \alpha \leq 180°$ in Theorem 5.2a is here superseded by $0 \leq 90° - \alpha \leq 180°$, which is the same as $-90° \leq \alpha \leq 90°$. Inequality (3) will be used primarily with acute angles α and β.

The device of replacing α and β by $90° - \alpha$ and $90° - \beta$, when applied to inequality (2) above gives

$$\sec \alpha + \sec \beta \geq 2 \sec \frac{\alpha + \beta}{2} \tag{4}$$

for $-90° < \alpha < 90°$, $-90° < \beta < 90°$, with equality iff $\alpha = \beta$.

Now we turn to the tangent and cotangent functions. First we prove that

$$\tan \alpha + \tan \beta \geq 2 \tan \frac{\alpha + \beta}{2} \tag{5}$$

for $0 \leq \alpha < 90°$, $0 \leq \beta < 90°$, with equality iff $\alpha = \beta$. There is equality in (5) if $\alpha = \beta$, so we presume that $\alpha \neq \beta$, say with $\alpha > \beta$, and prove strict inequality. For convenience write γ for $(\alpha + \beta)/2$, so that $\alpha - \gamma = \gamma - \beta$ and $\sin(\alpha - \gamma) = \sin(\gamma - \beta)$. This can be written as

$$\sin \alpha \cos \gamma - \sin \gamma \cos \alpha = \sin \gamma \cos \beta - \sin \beta \cos \gamma. \tag{6}$$

Also $\cos \alpha < \cos \beta$ since $\alpha > \beta$, and hence

$$(\cos \alpha \cos \gamma)^{-1} > (\cos \beta \cos \gamma)^{-1}. \tag{7}$$

Multiplying (6) and (7) we get

$$\tan\alpha - \tan\gamma > \tan\gamma - \tan\beta, \quad \text{or} \quad \tan\alpha + \tan\beta > 2\tan\gamma,$$

and this is (5) with strict inequality.

The replacement of α and β in (5) by $90° - \alpha$ and $90° - \beta$ gives the inequality

$$\cot\alpha + \cot\beta \geq 2\cot\frac{\alpha+\beta}{2} \tag{8}$$

for $0 < \alpha \leq 90°$, $0 < \beta \leq 90°$, with equality iff $\alpha = \beta$.

These inequalities give the background necessary to write an analog of Theorem 5.2b for the other trigonometric functions.

THEOREM 5.4a. *The inequality*

$$\cos\alpha_1 + \cos\alpha_2 + \cdots + \cos\alpha_n \leq n\cos\frac{\alpha_1+\alpha_2+\cdots+\alpha_n}{n} \tag{9}$$

holds if the α's satisfy $-90° \leq \alpha_j \leq 90°$, with equality iff the α's are all equal. The inequality

$$\tan\alpha_1 + \tan\alpha_2 + \cdots + \tan\alpha_n \geq n\tan\frac{\alpha_1+\alpha_2+\cdots+\alpha_n}{n} \tag{10}$$

holds if the α's satisfy $0 \leq \alpha_j < 90°$, with equality iff the α's are all equal. Results similar to (10) *hold for the secant, cosecant and cotangent functions with appropriate domains for the α's in each case. The domains are as indicated for the case $n = 2$ in* (4), (2) *and* (8) *above, respectively.*

E1. Prove that $\sin\alpha\sin\beta \leq \sin^2[(\alpha+\beta)/2]$ if $0 \leq \alpha \leq 180°$, $0 \leq \beta \leq 180°$, with equality iff $\alpha = \beta$. Prove a similar inequality for cosine in case $-90° \leq \alpha \leq 90°$, $-90 \leq \beta \leq 90°$.

E2. If α, β, γ, are the angles of any triangle, prove that the maximum of $\cos\alpha + \cos\beta + \cos\gamma$ is $3/2$, attained in the case of the equilateral triangle. Also prove that there is no minimum, but that the greatest lower bound is 1.

E3. Replace $\cos\alpha$ by $(b^2 + c^2 - a^2)/2bc$, and replace $\cos\beta$ and $\cos\gamma$ similarly, in the preceding problem, and so prove that

$$2abc < (a^2 + b^2 + c^2)(a + b + c) - 2(a^3 + b^3 + c^3) \leq 3abc$$

for the sides a, b, c of any triangle. (The restriction on a, b, c is important, as the case $a = b = 1$, $c = 3$ shows.)

E4. Find the maximum values of $\sin\alpha \sin\beta \sin\gamma$, and also of $\cos\alpha \cos\beta \cos\gamma$, for the angles α, β, γ of a triangle.

E5. Find the minimum value of $\operatorname{cosec}\alpha + \operatorname{cosec}\beta + \operatorname{cosec}\gamma$ for the angles of any triangle.

E6. For acute angles α and β prove that

$$\tan\alpha \tan\beta \leq \tan^2\frac{\alpha + \beta}{2} \qquad (*)$$

in case $\alpha + \beta < 90°$. If $90° < \alpha + \beta < 180°$, prove that the inequality $(*)$ is reversed, and that $(*)$ becomes an equality if $\alpha + \beta = 90°$. Except for the case $\alpha + \beta = 90°$, $(*)$ is an equality iff $\alpha = \beta$.

E7. If α, β, γ are the angles of a triangle, prove that

$$\tan\frac{\alpha}{2}\tan\frac{\beta}{2}\tan\frac{\gamma}{2} \leq (3\sqrt{3})^{-1},$$

with equality iff $\alpha = \beta = \gamma$. If the triangle is acute-angled, prove that $\tan\alpha \tan\beta \tan\gamma \geq 3\sqrt{3}$ and $\tan\alpha + \tan\beta + \tan\gamma \geq 3\sqrt{3}$, with equality iff $\alpha = \beta = \gamma$.

E8. Find the maximum value of $\sin x \sin y \sin(x + y)$ if $0 < x < 180°$, $0 < y < 180°$.

E9. Given two points P and Q on a circle, locate the point R on the circle so that the triangle PQR has (i) maximum area, (ii) maximum perimeter.

E10. Consider a triangle with sides a, b, c and opposite angles α, β, γ. Is it true that $a < (b + c)/2$ implies $\alpha < (\beta + \gamma)/2$? Is the converse true?

5.5. Extreme Values of $a \sin\theta + b \cos\theta$. For any constants a and b the maximum value of $a \sin\theta + b \cos\theta$ over all real values of θ is $\sqrt{a^2 + b^2}$, and the minimum is $-\sqrt{a^2 + b^2}$. This is easily proved by writing the expression in the form

$$\sqrt{a^2 + b^2}\left[\frac{a}{\sqrt{a^2 + b^2}}\sin\theta + \frac{b}{\sqrt{a^2 + b^2}}\cos\theta\right]. \tag{1}$$

For any real numbers a and b, not both zero, there is an angle λ satisfying

$$\sin\lambda = b/\sqrt{a^2 + b^2}, \qquad \cos\lambda = a/\sqrt{a^2 + b^2},$$

determined by the location of the point $(a/\sqrt{a^2 + b^2}, b/\sqrt{a^2 + b^2})$ on the unit circle $x^2 + y^2 = 1$. It follows that (1) can be written as

$$\sqrt{a^2 + b^2}\,[\cos\lambda \sin\theta + \sin\lambda \cos\theta]$$

$$= \sqrt{a^2 + b^2}\sin(\theta + \lambda).$$

The maximum is obtained by choosing θ so that $\sin(\theta + \lambda) = 1$, and the minimum so that $\sin(\theta + \lambda) = -1$.

If a and b are positive, the maximum value above can be obtained with a unique value of θ satisfying $0 \le \theta \le 90°$.

The procedure above can be extended to get the following result.

THEOREM 5.5a. *For any constant* $a > 1$, *the least value of* $a\sec\theta - \tan\theta$ *for* θ *satisfying* $0 \le \theta < 90°$ *is* $\sqrt{a^2 - 1}$, *occurring iff*

$$\sec\theta = a/\sqrt{a^2 - 1}, \qquad \tan\theta = 1/\sqrt{a^2 - 1}. \tag{2}$$

Furthermore, for any positive constant c the minimum value of

$$a\sqrt{c^2 + x^2} - x \tag{3}$$

over all real x is $c\sqrt{a^2 - 1}$, *occurring in case* $x = c/\sqrt{a^2 - 1}$.

To prove the first part of this theorem we write

$$m = a\sec\theta - \tan\theta; \tag{4}$$

so the problem is to minimize m. Now (4) gives $a = m\cos\theta + \sin\theta$. The expression on the right side of this equation resembles the linear function of $\sin\theta$ and $\cos\theta$ minimized earlier. This suggests that we define λ by

$$\cos\lambda = 1/\sqrt{1 + m^2}, \qquad \sin\lambda = m/\sqrt{1 + m^2},$$

with $0 < \lambda < 90°$. Hence $a = m \cos\theta + \sin\theta$ gives

$$a/\sqrt{1 + m^2} = \sin(\theta + \lambda).$$

To minimize m, we seek to maximize $a/\sqrt{1 + m^2}$. The maximum value of $\sin(\theta + \lambda)$ is 1, achieved by taking $\theta + \lambda = 90°$, and hence we have

$$a/\sqrt{1 + m^2} = 1, \quad m = \sqrt{a^2 - 1},$$

$$\sin\theta = \cos\lambda = 1/\sqrt{1 + m^2} = 1/a,$$

$$\cos\theta = \sin\lambda = m/\sqrt{1 + m^2} = \sqrt{a^2 - 1}/a.$$

These equations give the values of $\sec\theta$ and $\tan\theta$ listed in (2).

The second part of the theorem is a corollary of the first part. To minimize the expression (3) over all real x, we can restrict our attention to values of $x \geq 0$. The reason for this is that for any positive x, $F(-x) > F(x)$, where $F(x)$ is the expression (3).

For $x \geq 0$ define $\tan\theta = x/c$ with θ satisfying $0 \leq \theta < 90°$. Then $\sec\theta = \sqrt{(c^2 + x^2)}/c$, and the expression (3) is precisely

$$c[a \sec\theta - \tan\theta].$$

To minimize this we use the first part of the theorem, and this leads directly to

$$\tan\theta = 1/\sqrt{a^2 - 1}, \qquad x = c\tan\theta = c/\sqrt{a^2 - 1}.$$

Having completed the proof of the theorem, we note that the hypothesis $a > 1$ is essential, because for any other value of a the expression $a \sec\theta - \tan\theta$ has no minimum over values of θ satisfying $0 \leq \theta < 90°$. This is obvious in case $a \leq 0$, because $\tan\theta$ and $\sec\theta$ increase indefinitely as θ tends to $90°$. In case $0 < a \leq 1$, observe that

$$a \sec\theta - \tan\theta = \sec\theta - \tan\theta - (1 - a)\sec\theta$$

$$= \frac{1}{\sec\theta + \tan\theta} - (1 - a)\sec\theta$$

by using the identity $\sec^2\theta - \tan^2\theta = 1$. The fraction is positive, and approaches zero as θ tends to $90°$. Hence if $a = 1$ we conclude that $\sec\theta - \tan\theta$ has greatest lower bound 0 but no minimum. And if $0 < a < 1$ the term $(1 - a)\sec\theta$ in these equations shows that $a\sec\theta - \tan\theta$ becomes less than any prescribed real number as θ approaches 90°. Hence there is no minimum.

E11. Find the maximum value of $\sin\theta + \cos\theta$.

E12. Find the maximum of $4x + 3\sqrt{1 - x^2}$ over the domain $0 \le x \le 1$.

E13. Find the maximum of $4x + \sqrt{3 - x^2}$ over the domain $0 \le x \le \sqrt{3}$.

E14. Find the minimum of $5\sqrt{9 + 4x^2} - 8x$ over the domain of all real numbers x.

E15. If a and b are any given real numbers, find the maximum value of $a\cos^2\theta + b\sin^2\theta$ for all real numbers θ.

E16. Find the maximum value of each of: (a) $\sin\theta\sin 2\theta$; (b) $\sin\theta\cos 2\theta$; (c) $2\cos\theta + 3\sin^2\theta$.

E17. Among acute angles θ find the smallest value of $\tan\theta + 4\cot\theta$.

E18. Among angles θ satisfying $0 < \theta < 180°$, find the least value of $9\sin\theta + \operatorname{cosec}\theta$ and of $\sin\theta + 9\operatorname{cosec}\theta$.

5.6. Tacking Against a Headwind. *How should a sailboat tack against a north wind in order to go north as fast as possible?* As we proceed with the solution we describe this problem in greater detail, and in so doing we clarify what tacking is and how it is that a sailboat can move against the wind.

For convenience it is presumed that the wind is blowing steadily from the north, as indicated by the vector AB in Figure 5.6a. The boat moves in the direction indicated by BC, called the course of the boat. This direction can be chosen by the sailor, with the boat held to this course by the rudder and the keel. The sail is set in the direction BD, with the force of the wind transmitted via the sail to move the boat in the direction BC. The direction of the sail is also controlled by the sailor. The problem is to determine the appropriate directions BD and BC for the sail and the course, in order to achieve the fastest northerly motion of the boat.

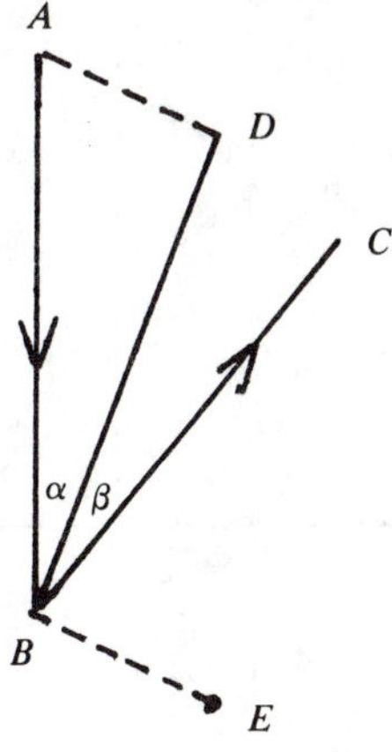

FIG. 5.6a

(As any reader familiar with sailing has already observed, certain aspects of the problem are oversimplified, so that actually we get only a first approximation to the solution. These simplifications will be discussed at the end of the section.)

One concept that is needed to analyze the problem is the idea of a *component* of a force or a velocity. Consider a force F indicated by the vector PQ in Figure 5.6b. The component of this force in the direction PT is given by the projection of PQ onto PT, namely, the vector PR, where QR is perpendicular to PT. The size or magnitude of this component is $F \cos \theta$ by simple trigonometry, where θ is the angle between the direction of the force and the direction of the component. A component of a velocity is defined in the same way.

In Figure 5.6a the angles ABD and DBC are denoted by α and β, respectively. If the force of the wind is denoted by F, then the thrust of the wind on the sail is the component of F in the direction AD, perpendicular to the sail. This component has magnitude

$$F \cos (90° - \alpha) = F \sin \alpha,$$

because the angle between AB and AD is $90° - \alpha$.

Now the force $F \sin \alpha$ has virtually no effect on the boat except to move it in the direction BC. So now we want to find the component of $F \sin \alpha$ in the direction BC. The force $F \sin \alpha$ in the direction AD

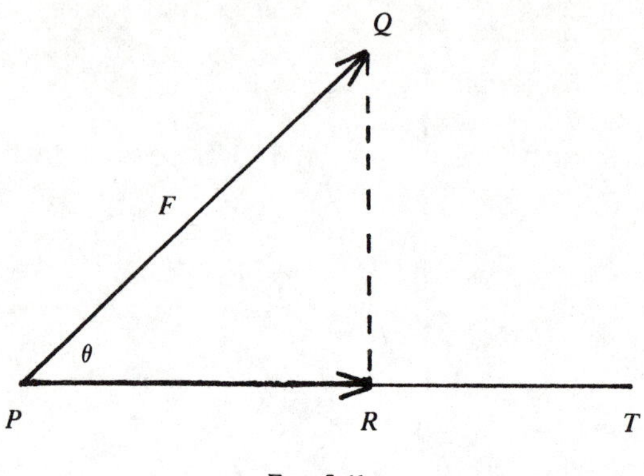

FIG. 5.6b

can be pictured just as well by BE, and the angle between this direction and the course direction BC is $90° - \beta$. Hence we multiply $F \sin \alpha$ by $\cos (90° - \beta)$ to get

$$F \sin \alpha \cos (90° - \beta) = F \sin \alpha \sin \beta.$$

This is the effective push of the wind on the boat, transmitted through the sail. The velocity of the boat along its course is proportional to this push. We presume a steady wind, so that F is a constant, and the velocity or speed of the boat in the direction BC is proportional to $\sin \alpha \sin \beta$. This argument shows that the sailboat can be moved "against" the wind, although not straight against it.

Now the problem is not the maximizing of $\sin \alpha \sin \beta$, but of the component of this velocity of the boat in the northerly direction, because it is the effective northbound motion of the boat that is to be as large as possible. The direction of motion is BC, and the angle ABC between this and the northerly direction is $\alpha + \beta$. Thus the northerly component of the velocity is proportional to

$$\sin \alpha \sin \beta \cos (\alpha + \beta).$$

Defining $\gamma = 90° - \alpha - \beta$, we see that the problem is to choose the angles α, β, γ so as to maximize $\sin \alpha \sin \beta \sin \gamma$, subject to the condition that $\alpha + \beta + \gamma = 90°$. By the result (9) of Section 5.3 the maximum is attained by taking $\alpha = \beta = \gamma = 30°$.

Thus the solution of the problem is this: To maximize the northerly component of the motion of the boat, take a course 60° *from due north, either* North 60° East, *or* North 60° West, *and set the sail midway between the direction of the wind and the direction of the course.*

Tacking means that to get from a point P to a second point T as shown in Figure 5.6c, a zigzag path is followed, for example along PQ, then QR, then RS, and finally ST. Because of the loss in speed when tacking at points Q, R and S, the fastest path from P to T would involve only one "tack," one zig and one zag. In actual sailing this is not practical because of water currents, shifting winds, and obstructions in the path.

The solution above is based on certain assumptions that are not fully realized in sailing. It is assumed that the sail offers a flat surface to the wind, whereas in actuality the sail, caught by the wind, is a curved surface with aerodynamic properties that cannot be analyzed by the simple components of forces used above. The effective angle α between the direction of the wind and the sail cannot be measured by simple observation; an experimental approach is needed.

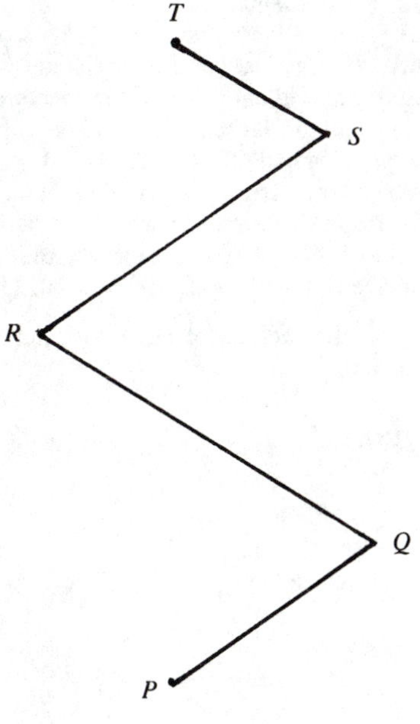

FIG. 5.6c

Furthermore, it is assumed above that the sailboat moves along a course in the direction of the keel, that is, in the direction the boat is pointed. We have not accounted for motion perpendicular to the keel caused by the component of the force of the wind in this perpendicular direction. The sailboat moves slightly crabwise through the water, with a small angle (called the angle of drift) between the course headed and the course actually sailed. When such additional factors are taken into account in the problem, the direction of course is noticeably less than 60° from the direction of the wind for maximum northerly velocity. Our mathematical analysis is, as we said, just a first approximation to the real situation.

Notes on Chapter 5

A discussion of Jensen's theorem with some further applications is given in the *Hungarian Problem Book II,* pp. 73-77. It should also be noted that both *Hungarian Problem Book I* and *Book II* contain many explanatory notes covering a wide variety of useful topics that are not always found in elementary textbooks on mathematics. These topics are conveniently indexed on page 109 of *Problem Book I* and page 118 of *Problem Book II.*

The basic inequalities that we use were formulated by J. L. W. V. Jensen (1906), who did much of the original work on the theory of convex functions. A comprehensive treatment of Jensen's inequalities in a broad setting is given in Hardy, Littlewood, and Pólya (1952, pp. 70-75, 84, 85).

Many of the results on the trigonometric functions of the angles of a triangle recur as problems in the popular journals. A sampling of such problems is given at the end of §5.4. Any reader interested in a wider variety of such questions should see the book by Bottema et al. (1968).

§5.6. The problem on the tacking of a sailboat has been adapted from Dörrie (1965, pp. 363-366).

CHAPTER 6

POLYGONS INSCRIBED AND CIRCUMSCRIBED

6.1. Introduction. Among all n-gons inscribed in a circle, which one has largest area and which one has largest perimeter? The regular polygon is the expected answer, and this turns out to be the case. On the other hand, among all n-gons circumscribed about a circle, the regular n-gon has the *least* area and the *least* perimeter. Similar questions for the ellipse are discussed in the next chapter.

The definitions of inscribed and circumscribed polygons are as follows: a polygon is said to be *inscribed* in a circle if every vertex of the polygon lies on the circle; a polygon is said to be *circumscribed* about a circle if every side of the polygon is a tangent line to the circle.

Another natural question is the comparison of regular polygons with n sides and with $n + 1$ sides. One would expect, again correctly, that the isoperimetric quotient increases from n to $n + 1$.

If the area and perimeter of a regular n-gon inscribed in a circle of radius 1 are denoted by A_n and L_n, then

$$A_n = \frac{1}{2}n \sin\frac{360^\circ}{n}, \qquad L_n = 2n \sin\frac{180^\circ}{n}. \tag{1}$$

These formulas are readily established by examining a triangle formed by the center of the circle and two adjacent vertices of the polygon, as illustrated in Figure 6.1a in the case $n = 6$.

The angles in formulas (1) are written in degree measure. The reason for this is that we want to use these results in §6.4 to outline a definition of π and the reasoning behind the circle formulas $C = 2\pi r$ and $A = \pi r^2$. Of course we must be careful not to use π in its

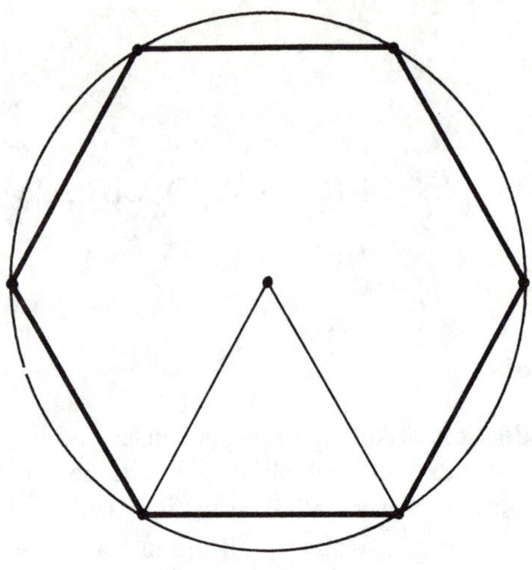

FIG. 6.1a

own definition. However, it will be convenient to use the radian measure π in place of 180° at other places in this chapter. But this can be regarded as just a notational usage for our purposes here, with π as a symbol to replace 180°. The intrinsic meaning of π and the circular formulas are not involved in any essential way.

Corresponding to formulas (1) we also need the area A_n' and the perimeter length L_n' of a regular polygon circumscribed about a unit circle; they are given by

$$A_n' = n \tan \frac{180°}{n}, \qquad L_n' = 2n \tan \frac{180°}{n}. \tag{2}$$

These results can be readily proved by examining a triangle formed by the center of the circle and two adjacent vertices of the polygon, as illustrated in Figure 6.1b in the case $n = 6$.

The work of this chapter could be simplified somewhat if the existence of maxima and minima is assumed. For example, if we assume that there is an n-gon of maximum area inscribed in a circle, it is fairly easy to conclude that it is the regular n-gon, by the following argument. We prove that if A, B, C are three consecutive vertices of the n-gon of maximum area, then $AB = BC$. It follows that all

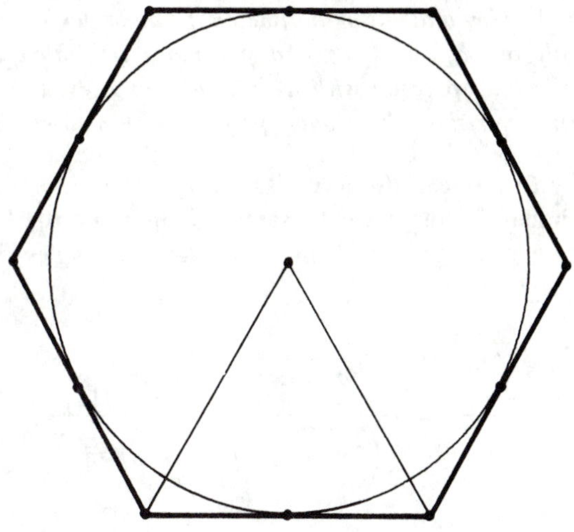

FIG. 6.1b

the sides are equal, and the n-gon is regular. For if $AB \neq BC$, we replace B by the midpoint B' of the arc ABC to get a triangle $AB'C$ of larger area than triangle ABC, because it has the same base AC but greater altitude.

In the special case $n = 4$, this argument can be adapted to prove, *without assuming the existence of a solution*, that among all quadrilaterals inscribed in a circle the square has the largest area. To see this, let $ABCD$ be any inscribed quadrilateral. If $ABCD$ is not a square, we show that the inscribed square has larger area. Choose B' as described in the preceding paragraph, the midpoint of the arc ABC. Similarly, choose D' as the midpoint of the arc ADC. Then $AB'CD'$ has larger area than $ABCD$, unless it coincides with $ABCD$. Note that $B'D'$ is a diameter of the circle. Next replace A and C by the midpoint A' of the arc $B'AD'$ and the midpoint C' of the arc $B'CD'$, respectively. Thus $A'B'C'D'$ is a square, with larger area than the quadrilateral $ABCD$.

6.2. Regular Polygons.

THEOREM 6.2a. *Among regular n-gons, the isoperimetric quotient $4\pi A/L^2$ increases with n. Stated otherwise, the regular polygon* *with $n + 1$ sides and given perimeter c has larger area than the regular polygon with n sides and perimeter c. Stated differently again, the regular polygon with $n + 1$ sides and given area k has a smaller perimeter than the regular polygon with n sides and area k.*

To prove this we can use formulas (2) from the preceding section, because similar figures have the same IQ. Ignoring the factor π in the isoperimetric quotient, and writing the angles in radian measure, we have

$$\frac{4A_n'}{(L_n')^2} = \frac{4n\tan\dfrac{\pi}{n}}{\left(2n\tan\dfrac{\pi}{n}\right)^2} = \frac{1}{n\tan\dfrac{\pi}{n}}. \tag{1}$$

To show that this increases with n amounts to proving that the denominator decreases. We prove that

$$n\tan\frac{\pi}{n} > (n+1)\tan\frac{\pi}{n+1} \tag{2}$$

for positive integers $n \geq 3$. This follows from inequality (10) in Theorem 5.4a with n replaced by $n + 1$, $\alpha_1 = 0$ and all the other angles $\alpha_2, \alpha_3, \ldots, \alpha_{n+1}$ equal to π/n. This gives

$$\tan 0 + n\tan\frac{\pi}{n} > (n+1)\tan\frac{\pi}{n+1},$$

with strict inequality because the angles are not all equal.

THEOREM 6.2b. *If A_n and L_n denote the area and perimeter of a regular n-gon inscribed in a circle of radius 1, then $A_n < A_{n+1}$ and $L_n < L_{n+1}$.*

By formulas (1) of the preceding section, $A_n < A_{n+1}$ amounts to

$$n\sin\frac{2\pi}{n} < (n+1)\sin\frac{2\pi}{n+1}. \tag{3}$$

To prove this we use Theorem 5.2b with $n + 1$ angles

$$\alpha_1 = 0, \quad \alpha_2 = 2\pi/n, \quad \alpha_3 = 2\pi/n, \quad \ldots, \quad \alpha_{n+1} = 2\pi/n$$

with arithmetic mean $2\pi/(n + 1)$. The outcome is inequality (3) because $\sin 0 = 0$.

The proof that $L_n < L_{n+1}$ is very similar, because by formulas (1) of the preceding section what is to be proved amounts to

$$n \sin\frac{\pi}{n} < (n + 1)\sin\frac{\pi}{n + 1}.$$

This is a direct consequence of Theorem 5.2b applied to the $n + 1$ angles $0, \pi/n, \pi/n, \ldots, \pi/n$.

THEOREM 6.2c. *If A_n' and L_n' denote the area and perimeter of a regular n-gon circumscribed about a circle of radius* 1, *then $A_n' > A_{n+1}'$ and $L_n' > L_{n+1}'$.*

Using the formulas (2) from the preceding section we see that these results amount to the same thing, namely, the inequality

$$n \tan\frac{\pi}{n} > (n + 1)\tan\frac{\pi}{n + 1}.$$

This was established above as inequality (2).

6.3. Inscribed and Circumscribed Polygons.

THEOREM 6.3a. *Among all triangles inscribed in a circle, the equilateral triangle has the largest area and the largest perimeter.*

We note that it suffices to prove this for circles of radius 1, because, given any circle, we can take its radius as the unit length. First consider any inscribed triangle T (not equilateral) such that the center of the circle lies inside or on T. Then if the three sides subtend angles α, β, γ at the center, we see that $\alpha + \beta + \gamma = 360°$. Also the area and perimeter of T are readily calculated to be

$$\frac{1}{2}(\sin \alpha + \sin \beta + \sin \gamma) \quad \text{and} \quad 2(\sin \alpha/2 + \sin \beta/2 + \sin \gamma/2).$$

By Theorem 5.2b we conclude that each of these expressions is a maximum iff $\alpha = \beta = \gamma$. Thus the equilateral triangle has larger area and perimeter than T.

Second, consider any triangle inscribed in the unit circle such that the center of the circle lies outside the triangle. If the two smaller sides of such a triangle subtend angles α, β at the center of the circle, then the largest side subtends an angle $\alpha + \beta$, as shown in Figure 6.3a. Thus the area of the triangle PQR in the figure is

$$\begin{aligned} \frac{1}{2}\sin\alpha + \frac{1}{2}\sin\beta - \frac{1}{2}\sin(\alpha+\beta) &< \frac{1}{2}\sin\alpha + \frac{1}{2}\sin\beta \\ &\leq \sin[(\alpha+\beta)/2] \\ &< \sin 90^\circ = 1. \end{aligned}$$

But this is less than the area $3\sqrt{3}/4$ of an equilateral triangle inscribed in the unit circle.

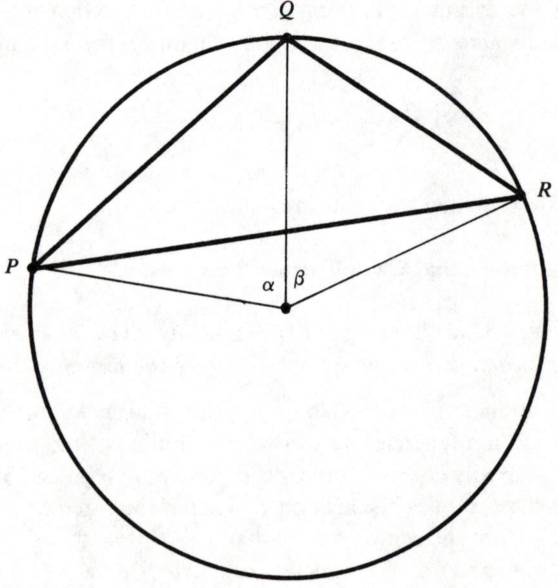

FIG. 6.3a

The perimeter of this triangle PQR is readily calculated to be

$$\begin{aligned} &2\sin\frac{\alpha}{2} + 2\sin\frac{\beta}{2} + 2\sin\frac{\alpha+\beta}{2} \\ &\leq 4\sin\frac{\alpha+\beta}{4} + 2\sin\frac{\alpha+\beta}{2} \\ &< 4\sin 45^\circ + 2\sin 90^\circ = 2\sqrt{2} + 2. \end{aligned}$$

This is less than $3\sqrt{3}$, the perimeter of an equilateral triangle inscribed in a unit circle.

THEOREM 6.3b. *Among all n-gons inscribed in a circle, the regular n-gon has the largest area and the largest perimeter.*

Since this result includes Theorem 6.3a as the special case $n = 3$, we first explain why we prove two separate theorems here. The reason is that the triangle case, $n = 3$, is more complicated than the general case because of the awkwardness involved in triangles like PQR in Figure 6.3a. In the general case the matter is simpler.

To prove Theorem 6.3b, we observe again that there is no loss of generality in taking the radius of the circle as 1. First consider the case of an inscribed n-gon P (not regular) such that the center of the circle lies inside or on the n-gon. Let $\alpha_1, \alpha_2, \ldots, \alpha_n$ be the angles subtended at the center of the circle by the sides of the n-gon P, so that

$$\alpha_1 + \alpha_2 + \cdots + \alpha_n = 360^\circ.$$

The angles are not all equal, since the polygon is not regular. The area of this n-gon is

$$\begin{aligned} &\frac{1}{2}[\sin\alpha_1 + \sin\alpha_2 + \cdots + \sin\alpha_n] \\ &\qquad < \frac{n}{2}\sin\frac{\alpha_1+\alpha_2+\cdots+\alpha_n}{n} = \frac{n}{2}\sin\frac{360^\circ}{n}, \end{aligned}$$

where the inequality comes from a direct application of Theorem 5.2b, because each angle α_j satisfies $0 < \alpha_j \leq 180^\circ$. Hence the regular n-gon has larger area, in view of formulas (1) of §6.1.

The perimeter of the n-gon P described above is

$$2\left[\sin\frac{\alpha_1}{2} + \sin\frac{\alpha_2}{2} + \cdots + \sin\frac{\alpha_n}{2}\right]$$
$$< 2n \sin\frac{\alpha_1 + \alpha_2 + \cdots + \alpha_n}{2n} = 2n \sin\frac{180^\circ}{n}.$$

Thus the regular n-gon has larger perimeter.

To complete the proof of Theorem 6.3b we must consider the second case, where the center of the unit circle is outside the inscribed n-gon P. In this case we prove again that the area and perimeter of P are less than the area and perimeter of a regular inscribed n-gon. By Theorem 6.2b and formulas (1) of Section 6.1 with $n = 4$, we see that the area and perimeter of a regular n-gon inscribed in the unit circle are at least 2 and $4\sqrt{2}$, respectively, provided $n > 3$. We can take $n > 3$ because the triangle was treated specially in Theorem 6.3a. On the other hand, the inscribed polygon P lies completely inside a semicircle of radius 1, and it is not difficult to see that the area and perimeter of P are less than the area and perimeter, respectively, of the semicircle. The area of the semicircle is $\pi/2$, and the perimeter is $2 + \pi$. The argument is completed by observing that $2 > \pi/2$ and $4\sqrt{2} > 2 + \pi$.

THEOREM 6.3c. *Among all n-gons circumscribing a circle, the regular n-gon has the smallest area and the smallest perimeter, but the largest isoperimetric quotient.*

Once again it may be assumed without loss of generality that the circle has radius 1. Consider any nonregular n-gon P that is circumscribed about the circle. There are n points of tangency of the sides of the n-gon with the circle, as illustrated in Figure 6.3b in the case $n = 5$. Drawing the radii to these points of tangency, we have n angles at the center of the circle. Label these angles $2\alpha_1$, $2\alpha_2$, $\ldots$, $2\alpha_n$. Note that $2\alpha_j < 180^\circ$ and so $\alpha_j < 90^\circ$ for $j = 1, 2, \ldots, n$, and that

$$\alpha_1 + \alpha_2 + \cdots + \alpha_n = 180^\circ.$$

The n-gon is separated by the radii to the points of tangency into n quadrilaterals, the lengths of whose sides and whose areas are easy

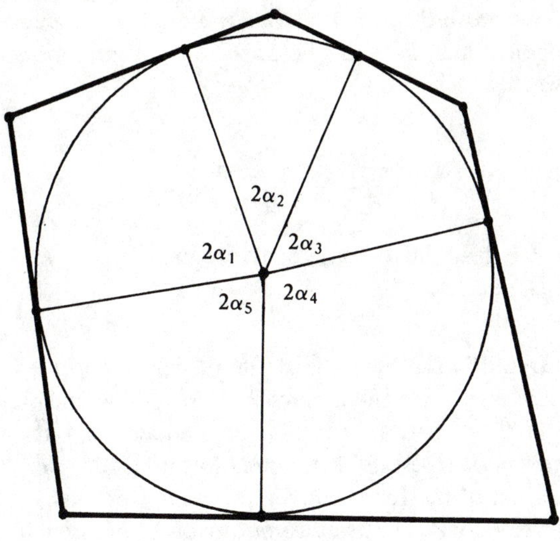

FIG. 6.3b

to calculate. The area A and the perimeter L of the n-gon P satisfy the relations

$$A = \tan\alpha_1 + \tan\alpha_2 + \cdots + \tan\alpha_n, \qquad L = 2A. \tag{1}$$

Since the n-gon is not regular, the angles $\alpha_1, \alpha_2, \ldots, \alpha_n$ are not all equal. Hence by Theorem 5.4a we see that the area of the n-gon P satisfies

$$\begin{aligned} &\tan\alpha_1 + \tan\alpha_2 + \cdots + \tan\alpha_n \\ &> n\tan\frac{\alpha_1 + \alpha_2 + \cdots + \alpha_n}{n} = n\tan\frac{180^\circ}{n}. \end{aligned} \tag{2}$$

By formulas (2) of §6.1, we know that $n\tan(180^\circ/n)$ is the area of the regular n-gon circumscribed about the unit circle. Thus we have proved that the regular n-gon has smaller area than any other circumscribed n-gon P.

The same result holds for the perimeters, because the equation $L = 2A$ holds for the perimeter L and the area A of any polygon circumscribed about a unit circle.

Finally, we prove that the isoperimetric quotient is greatest for the regular n-gon. Since $L = 2A$ holds for any circumscribed polygon, we observe that

$$\frac{4\pi A}{L^2} = \frac{4\pi A}{4A^2} = \frac{\pi}{A}.$$

The area A is least in the case of the regular n-gon, and it follows that π/A is greatest.

6.4. A Definition of π. The standard definition of π is *the ratio of the circumference to the diameter in any circle*, corresponding to the formula $C = 2\pi r$. In addition to spelling this out more completely, we shall also establish the area formula $A = \pi r^2$. All of this is a by-product of the results of the preceding sections.

In order to be sure that we are not involved in circular reasoning by using some hidden definition of π in a covert way, let us outline the presumed background. We need to have the basic concepts of the length of a line segment and the area of a triangle. Clearly π is not involved in these. Also we need the trigonometric inequalities of the preceding chapter. All this can be based on trigonometry using *degree* measure of angles. Radian measure follows the definition of π and the formula $C = 2\pi r$. The material in §6.1 and §6.2 is also used. Note that the only aspect of a circle needed is its definition as the set of points in the plane at a fixed distance from a fixed point.

Also needed is the simple idea of an infinite sequence of nested intervals $I_1, I_2, I_3, \ldots$ on the real line, where each interval is contained in the preceding one. If the length of I_n tends to zero as n tends to infinity, then we assume that there is a unique real number common to all the intervals. This idea has already been introduced in §2.8, where the number e was defined by taking I_n to be the interval from $f(n)$ to $g(n)$ as outlined there.

As in §6.1, A_n and L_n denote the area and perimeter of a regular n-gon inscribed in a circle of radius 1, and A_n' and L_n' denote the area and perimeter of a circumscribed regular n-gon. By Theorems 6.2b and 6.2c the inequalities

$$A_n < A_{n+1}, \quad L_n < L_{n+1}, \quad A_n' > A_{n+1}', \quad L_n' > L_{n+1}' \qquad (1)$$

hold for every integer $n \geq 3$. It is helpful also to recognize that

$$A_n < A_k', \qquad L_n < L_k' \tag{2}$$

for all pairs n, k of positive integers ≥ 3. The first of these is obvious, because A_n is the area of a polygon contained in the circle, and A_k' is the area of a polygon containing the circle.

To prove the second part of (2), we start by establishing $L_n < L_n'$. From formulas (1) and (2) of §6.1 we have

$$\begin{aligned} L_n' - L_n &= 2n \tan \frac{180°}{n} - 2n \sin \frac{180°}{n} \\ &= 2n \tan \frac{180°}{n} \left[1 - \cos \frac{180°}{n}\right], \end{aligned}$$

and this is positive, so $L_n' > L_n$. If $n > k$ it is clear that $L_n < L_n' < L_k'$. On the other hand if $n < k$ we can write $L_n < L_k < L_k'$, and hence the proof of (2) is complete.

The equations above for $L_n' - L_n$ reveal something more, namely, that $L_n' - L_n$ tends to zero as n tends to infinity. Since $L_4' > L_n'$ for $n > 5$, and since $L_4' = 8$, we have

$$L_n' - L_n = L_n'\left(1 - \cos \frac{180°}{n}\right) < L_4'\left(1 - \cos \frac{180°}{n}\right) = 8\left(1 - \cos \frac{180°}{n}\right).$$

As n increases, $\cos(180°/n)$ approaches $\cos 0$, which is 1. Hence $1 - \cos(180°/n)$ tends to zero as n tends to infinity.

Thus if we define the interval I_n to be (L_n, L_n'), that is, the open interval from L_n to L_n', we have a set of nested intervals $I_3, I_4, I_5, \ldots$. As n tends to infinity, the lengths of these intervals tend to zero, and so *they define a unique real number*. Because the circumference of the circle lies between the polygons with perimeters L_n and L_n', it is natural to say that this real number is the length of the circumference of a circle of radius 1, or diameter 2, and so we denote it by 2π.

For a circle of radius r, all lengths are proportionally larger, or smaller, by the factor r. In this case the perimeter of the inscribed regular n-gon is rL_n, and that of the circumscribed regular n-gon is rL_n'. The nested intervals (rL_n, rL_n') contain the unique number $2\pi r$, and this is the length of the circumference. Thus we obtain the formula $C = 2\pi r$.

For a circle of radius 1, the difference $A_n' - A_n$ between the areas of the circumscribed and inscribed regular n-gons also tends to zero as n tends to infinity. To prove this, note that

$$\begin{aligned} A_n' - A_n &= n \tan \frac{180°}{n} - \frac{1}{2} n \sin \frac{360°}{n} \\ &= n \tan \frac{180°}{n} - n \sin \frac{180°}{n} \cos \frac{180°}{n} \\ &= n \tan \frac{180°}{n} \left[1 - \cos^2 \frac{180°}{n} \right] \\ &= A_n' \left[1 - \cos^2 \frac{180°}{n} \right] \\ &< A_4' \left[1 - \cos^2 \frac{180°}{n} \right] \quad \text{for } n > 4. \end{aligned}$$

The term in square brackets tends to zero as n gets large, and $A_4' = 4$.

So again we have a set of nested intervals

$$(A_3, A_3'), (A_4, A_4'), (A_5, A_5'), \ldots, (A_n, A_n'), \ldots \tag{3}$$

with lengths tending to zero. To what number do these intervals converge? The earlier sequence of intervals (L_n, L_n') converged to 2π. Consequently, the sequence of intervals $(\frac{1}{2}L_n, \frac{1}{2}L_n')$ would converge to π itself. Since $\frac{1}{2}L_n' = A_n'$, this sequence can be written as $(\frac{1}{2}L_n, A_n')$. Having the same upper endpoint A_n' as the sequence (3), this sequence must converge to the same limit, which is π as we have just seen. The area of a circle of radius 1 is therefore π.

To get the area of a circle of radius r, we note that the area of the inscribed regular n-gon is $r^2 A_n$, and the area of the circumscribed regular n-gon is $r^2 A_n'$. Thus all areas are multiplied by r^2 as we pass from radius 1 to radius r, and we get the formula $A = \pi r^2$ for the area of a circle of radius r.

6.5. Circles Versus Regular Polygons. In the short table of isoperimetric quotients in §4.4, the circle has a larger value than the regular octagon, hexagon, or pentagon. This is to be expected in view of Theorem 4.3b, the isoperimetric theorem. However, that theorem is proved with the existence of a solution assumed. We now give an independent proof, without any existence assumption, that the IQ of any regular n-gon is less than that of a circle. (This result is proved for *any* n-gon in Chapter 12.)

THEOREM 6.5a. *If θ is any acute angle measured in radians, then $\theta < \tan\theta$. It follows that the isoperimetric quotient of any regular polygon is less than that of a circle.*

We begin by proving that the second part follows from the first. For a circle, the isoperimetric quotient $4\pi A/L^2$ is 1. For a regular n-gon, the isoperimetric quotient is

$$\frac{4\pi n \tan\dfrac{\pi}{n}}{\left(2n\tan\dfrac{\pi}{n}\right)^2} = \frac{\dfrac{\pi}{n}}{\tan\dfrac{\pi}{n}}$$

by use of formulas (2) of §6.1 in radian measure form. The last fraction is less than 1 in view of the first part of the theorem, $\theta < \tan\theta$.

To prove the first part consider the sector CPQ of the unit circle as shown in Figure 6.5a, with the arc PQ subtending an angle θ at the center C. Since the radius is 1, the area of the entire circle (not shown in the diagram) is π; and hence the area of the sector CPQ is the proportional part with an angle θ at the center. Thus the area of the sector CPQ is $\theta/2$ from the calculation $(\theta/2\pi)\cdot\pi$. Let R be the point of intersection of the line CQ extended and the perpendicular drawn to CP at P. (Thus the line PR is tangent to the circular arc at P.) Since $CP = 1$, the length of PR is $\tan\theta$, so the area of triangle CPR is $\frac{1}{2}\tan\theta$. Comparing the areas of the sector CPQ and the triangle CPR we get

$$\frac{1}{2}\theta < \frac{1}{2}\tan\theta \quad \text{or} \quad \theta < \tan\theta.$$

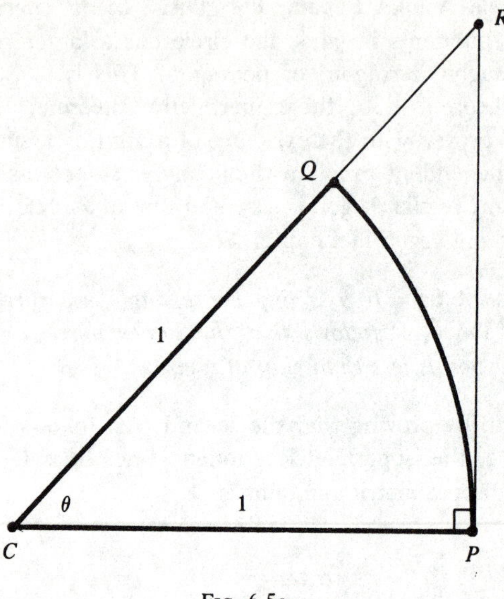

FIG. 6.5a

F1. Prove that if θ is any acute angle, measured in radians, then

$$\text{(i)}\quad \sin\theta < \theta, \qquad \text{and} \qquad \text{(ii)}\quad \cos\theta > 1 - \theta.$$

Suggestion for part (i): In Figure 6.5a compare the areas of the triangle CPQ and the sector CPQ.

F2. Consider the set of all quadrilaterals inscribed in a semicircle, with two vertices at the ends of the diameter. Which has maximum area?

F3. Find the maximum value of each of $x + y$, xy, and $3x + 4y$ if x and y are restricted to $x^2 + y^2 = 1$.

F4. For any integer $n \geq 2$ prove that n not necessarily distinct points can be chosen on the unit circle so that the sum of the squares of all distances between pairs of points is n^2, but that no sum larger than n^2 is possible. (Suggestion: Use a coordinate system with the circle $x^2 + y^2 = 1$.) For the extension of this question to a sphere, with a slightly surprising outcome, see Problem K15 of Section 11.4.

F5. In Figure 6.5a prove that the length of the line segment PR exceeds the length of the arc PQ. Hence prove that any polygon circumscribed about a circle has greater perimeter than the circle. (Another way to prove this result is to establish that $L > 2\pi r$, where L is the perimeter of a polygon circumscribed about a circle of radius r. To do this, observe that the area A of such a polygon satisfies $A = rL/2$ and $A > \pi r^2$; see Figure 6.3b.)

Notes on Chapter 6

The approach to inscribed and circumscribed polygons in §6.3 is unified by the use of the trigonometric inequalities of the preceding chapter. For geometric treatments of the inscribed n-gon of maximum area, see Benson (1966, p. 131), Levenson (1967, p. 134), and Pólya (1954, p. 127). Geometric approaches often find the question of existence a bothersome one, as the Levenson and Pólya references show.

Fejes Tóth (1947) has given a direct geometric proof, elementary but sophisticated, of the "least area" part of Theorem 6.3c. A very clear formulation of this can be found in Honsberger (1976, pp. 13–16).

§6.4. For an interesting discussion of the area of a circle, see Epstein and Hochberg (1977).

§6.5. Problem F5 is a special case of this result: if a convex region R_1 lies inside a convex region R_2, the perimeter of R_2 exceeds that of R_1.

CHAPTER 7

ELLIPSES

7.1. A Basic Mapping. Any ellipse can be represented by an equation

$$\frac{x^2}{a^2} + \frac{y^2}{b^2} = 1 \tag{1}$$

by a proper choice of a system of axes, where a and b are positive constants. It is conventional to take $a > b$. The graph of (1), illustrated in Figure 7.1a, is an example of a *centrally symmetric* curve, which means that every straight line through the center cuts the curve at two points equally distant from the center. The center of (1) is the origin of the coordinate system. The major axis is the line segment from $(-a,0)$ to $(a,0)$, whereas the minor axis extends from $(0,-b)$ to $(0,b)$.

Many problems about the ellipse can be solved by reduction to a circle. One way to do this is by use of the mapping, or transformation,

$$x = aX, \quad y = bY, \tag{2}$$

which reduces equation (1) to $X^2 + Y^2 = 1$. This is a unit circle in the new system of coordinates X, Y. Every point (x, y) in the first system of coordinates is mapped onto a point (X, Y) in the new system. Every straight line $cx + dy + e = 0$ is mapped onto a straight line $caX + dbY + e = 0$ in the new coordinate system.

One of the most useful properties of the mapping (2) is the simple relation that holds between the area of a region in the xy plane and the area of the corresponding region in the XY plane. Specifically,

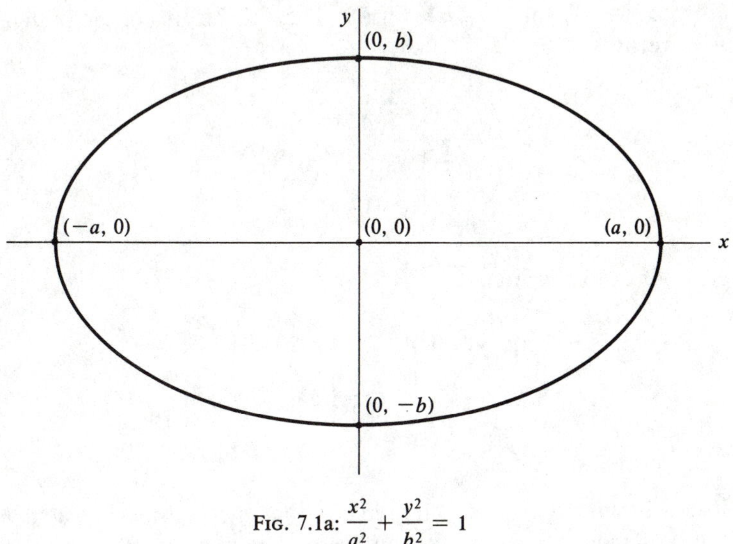

FIG. 7.1a: $\dfrac{x^2}{a^2} + \dfrac{y^2}{b^2} = 1$

consider a region R having an area A in the xy plane which maps onto a region R' with area A' in the XY plane. Then it is easy to prove that

$$A = abA' \quad \text{or} \quad A' = A/ab. \tag{3}$$

To verify this, we assume the reader is familiar with the fact that the area of the triangle with vertices (x_1, y_1), (x_2, y_2), (x_3, y_3) is given by the determinant formula

$$\frac{1}{2}\begin{vmatrix} x_1 & y_1 & 1 \\ x_2 & y_2 & 1 \\ x_3 & y_3 & 1 \end{vmatrix} = \frac{1}{2}(x_1y_2 + x_2y_3 + x_3y_1 - x_1y_3 - x_2y_1 - x_3y_2), \tag{4}$$

provided the three vertices are listed in counterclockwise order around the triangle. (If the points are listed in clockwise order, then formula (4) gives the negative of the area.)

The triangle with vertices (x_j, y_j) for $j = 1, 2, 3$ is mapped onto the triangle with vertices

$$(X_1, Y_1), (X_2, Y_2), (X_3, Y_3), \tag{5}$$

where $x_i = aX_i$ and $y_i = bY_i$ for $i = 1, 2, 3$. The area of the triangle with vertices (5) is

$$\frac{1}{2}\begin{vmatrix} X_1 & Y_1 & 1 \\ X_2 & Y_2 & 1 \\ X_3 & Y_3 & 1 \end{vmatrix}. \tag{6}$$

Now (4) can be written

$$\frac{1}{2}\begin{vmatrix} x_1 & y_1 & 1 \\ x_2 & y_2 & 1 \\ x_3 & y_3 & 1 \end{vmatrix} = \frac{1}{2}\begin{vmatrix} aX_1 & bY_1 & 1 \\ aX_2 & bY_2 & 1 \\ aX_3 & bY_3 & 1 \end{vmatrix}.$$

It is a basic property of determinants that the multipliers a and b can be removed from the first and second columns thus,

$$\frac{1}{2}\begin{vmatrix} aX_1 & bY_1 & 1 \\ aX_2 & bY_2 & 1 \\ aX_3 & bY_3 & 1 \end{vmatrix} = \frac{ab}{2}\begin{vmatrix} X_1 & Y_1 & 1 \\ X_2 & Y_2 & 1 \\ X_3 & Y_3 & 1 \end{vmatrix}.$$

This proves the relation $A = abA'$ in the case of triangles.

Thus the property (3) holds for triangles, and so also for any figure that can be triangulated, i.e., separated into triangles. Hence (3) holds for the area A of a polygon P in the xy plane and the area A' of its image in the XY plane. Now the area of a region bounded by a curve is defined by some scheme involving polygons that approximate the region, for example, by the process sketched briefly in §4.1. Hence the property (3) passes over from polygons to regions in general, that is, regions for which area can be defined.

We can use this result to get the area of the ellipse from the area of the circle. That is, the result $A = abA'$ holds in case A denotes the area enclosed by the ellipse (1), where A' denotes the area enclosed by the circle $X^2 + Y^2 = 1$. Now this is the unit circle, i.e., the circle with radius 1, so its area is π. Thus $A' = \pi$ and so $A = ab\pi$. So we have deduced the standard formula that the area enclosed by the ellipse (1) is πab.

The relationship $A = abA'$ between areas has no analog for the lengths of straight line segments. The following example shows that for lengths there is no simple relationship. The line segment from $(0, 0)$ to $(a, 0)$ in the (x, y) coordinate system has length a. The mapping (2) carries these points into $(0, 0)$ and $(1, 0)$, giving a line segment of length 1. This suggests the simple relation $L = aL'$ between a length L in the first coordinate system to the corresponding length L' in the second.

However, consider also the line segment from $(0, 0)$ to $(0, b)$ in the (x, y) system, a segment of length b. The mapping (2) carries these two points into $(0, 0)$ and $(0, 1)$, giving a line segment of length 1. Thus we have a segment of length b mapped onto a segment of length 1, whereas the preceding example mapped a segment of length a onto a segment of length 1. Since $a \neq b$, there is no relation between lengths as simple as the area relationship $A = abA'$.

G1. Assuming that $a > b > 0$, does the mapping $x = aX$, $y = bY$ carry parallel lines in the xy plane into parallel lines in the XY plane? Does the mapping carry perpendicular lines into perpendicular lines?

G2. If four points on an ellipse form a rectangle, prove that the sides of the rectangle are parallel to the axes of the ellipse.

7.2. Parametric Equations. The equations

$$x = a\cos\theta, \qquad y = b\sin\theta \tag{1}$$

are called *parametric equations* of the ellipse

$$\frac{x^2}{a^2} + \frac{y^2}{b^2} = 1 \tag{2}$$

for the following reason. If the equations (1) are divided by a and b, respectively, and then squared and added, the outcome is (2) by virtue of the identity $\cos^2\theta + \sin^2\theta = 1$, which is just the theorem of Pythagoras written in trigonometric form. In fact as θ runs through all values from $\theta = 0$ to $\theta = 2\pi$ radians, the points (x, y) given by equations (1) are precisely the points on the ellipse (2). For example $\theta = 0$ and $\theta = 2\pi$ both give the point $(a, 0)$, and $\theta = \pi/2$ gives the point $(0, b)$. The variable θ is called a *parameter*, and hence we have the name "parametric equations."

It should be noted (for readers familiar with polar coordinates) that θ does not have the same meaning as it has in the system of polar coordinates. For example, if we take $\theta = \pi/4$ or $\theta = 45°$ in (1) we get the point $(a/\sqrt{2}, b/\sqrt{2})$. Since $a \neq b$, these coordinates are not equal, and hence this is not the point where the 45° line through the origin, namely, the line $y = x$, intersects the ellipse in the first quadrant.

7.3. Polygons Inscribed in an Ellipse. A standard problem in calculus books asks for the rectangle of largest area that can be inscribed in the ellipse $x^2/a^2 + y^2/b^2 = 1$. Stated in this way, the question is not too easy, because it involves a consideration of all possible rectangles that can be inscribed. Specifically, are there any rectangles with vertices on the given ellipse whose sides are not parallel to the coordinate axes? The answer is no, as we have seen in problem G2 in §7.1. To avoid this question of precisely which rectangles can be inscribed, the more carefully written calculus books include some such qualifying phrase as "among rectangles with sides parallel to the axes."

Our purpose in this section is to solve *the more general problem of finding four points on an ellipse to give a quadrilateral of maximum area.* There are infinitely many such quadrilaterals, one of which is the rectangle asked for in the calculus books. Even more generally, we solve the problem of finding n points on an ellipse to give an n-gon of maximum area.

THEOREM 7.3a. *Among all quadrilaterals inscribed in the ellipse $x^2/a^2 + y^2/b^2 = 1$, there are infinitely many having the largest possible area, 2ab. The quadrilaterals have vertices*

$$(\pm a \cos\theta, \pm b \sin\theta), \quad (\pm a \sin\theta, \mp b \cos\theta), \tag{1}$$

where θ satisfies $0 \le \theta < \pi/2$, and where in each pair of points it is understood that the upper signs are taken together, and likewise the lower signs.

Remark: This result could be stated with θ allowed to be an arbitrary value. The reason for restricting θ by the inequalities $0 \leq \theta < \pi/2$ is to avoid repetitions of answers. For example, the four points (1) given by $\theta = 0$ are $(\pm a, 0)$ and $(0, \pm b)$. These same points are obtained by taking $\theta = \pi/2$, $\theta = \pi$, and $\theta = 3\pi/2$.

If we take $\theta = \pi/4$, the four points (1) are

$$(\pm \mathrm{a}/\sqrt{2}, \pm b/\sqrt{2}), \quad (\pm a/\sqrt{2}, \mp b/\sqrt{2}). \tag{2}$$

These four points form a rectangle, and this is the only rectangle among the quadrilaterals (1). (The proof of this is asked in a problem at the end of this section.)

The proof of Theorem 7.3a is very simple. Using the mapping $x = aX$, $y = bY$ of §7.1, the ellipse is transformed into the circle $X^2 + Y^2 = 1$. Any four points P, Q, R, S on the ellipse are mapped onto four points P', Q', R', S' on the circle. The area A of the quadrilateral $PQRS$ and the area A' of the quadrilateral $P'Q'R'S'$ are related by the formula $A = abA'$, as given by formula (3) of §7.1.

So in order to maximize the area A, we seek the maximum value of area A'. By Theorem 6.3b the area A' of a quadrilateral $P'Q'R'S'$ inscribed in the circle $X^2 + Y^2 = 1$ is maximized if the quadrilateral is a square. This can be achieved by taking P' in arbitrary fashion with coordinates $(\cos\theta, \sin\theta)$ and then taking Q', R', and S' as

$$(-\sin\theta, \cos\theta), \quad (-\cos\theta, -\sin\theta), \quad \text{and} \quad (\sin\theta, -\cos\theta)$$

equally spaced around the circle from P' to form a square. Although P' is arbitrary, we can avoid duplicate answers here by limiting θ to $0 \leq \theta < \pi/2$; thus P' is chosen on the X-axis or in the first quadrant.

The mapping $x = aX$, $y = bY$ then leads us to points (x, y) with coordinates precisely as described in formulas (1) in the statement of the theorem.

Thus far in this section we have discussed quadrilaterals of maximum area inscribed in an ellipse. The method clearly extends to polygons of any number of sides, because of the reduction of the problem to a circle. Here is the general result:

THEOREM 7.3b. *Among all n-gons inscribed in the ellipse, $x^2/a^2 + y^2/b^2 = 1$, there are infinitely many with maximum area. In fact, given any point P on the ellipse, there is an n-gon of maximum area having P as one of its vertices. If P has coordinates* $(a \cos \theta, b \sin \theta)$ *then all the vertices of the n-gon are*

$$(a \cos(\theta + 2k\pi/n), b \sin(\theta + 2k\pi/n)), \tag{3}$$

where $k = 0, 1, 2, 3, \ldots, n - 1$.

If $a = b = 1$, the points (3) are the vertices of a regular n-gon inscribed in the unit circle. The proof of Theorem 7.3b is the natural extension of the argument in Theorem 7.2a from four points to n points, and so the details are omitted.

G3. Prove that the four points in formula (1) form a rectangle in case $\theta = \pi/4$ and that the points form a parallelogram for all other values of θ satisfying $0 \leq \theta < \pi/2$.

G4. What is the area of a triangle of largest area inscribed in $x^2/a^2 + y^2/b^2 = 1$? What is the area of the n-gon of largest area?

7.4. Circumscribed Polygons. The question of a polygon of smallest area circumscribed about the ellipse $x^2/a^2 + y^2/b^2 = 1$ can also be reduced to the corresponding problem for a circle.

The additional information needed to solve the problem is this: If the mapping $x = aX$, $y = bY$ is applied to the ellipse and to any line tangent to the ellipse, the outcome is the circle $X^2 + Y^2 = 1$ and a line tangent to it.

The reason for this is inherent in the nature of a tangent line. Any line drawn on the same plane as an ellipse has exactly one of three properties, namely, that it intersects the ellipse at two, one, or no points. The lines that have one and only one point in common with the ellipse are the tangent lines. Consequently, the mapping $x = aX$, $y = bY$ applied to the ellipse $x^2/a^2 + y^2/b^2 = 1$ and a tangent line transform them into a circle and a line with one and only one point in common with the circle, that is, a tangent line to the circle.

It follows that the problem of finding n-gons of smallest area circumscribed about an ellipse reduces to the question of n-gons of minimum area circumscribed about a circle. This question was dealt with in Theorem 6.3c, and so we have the following extension.

THEOREM 7.4a. *Let n be a fixed positive integer ≥ 3. Among all n-gons circumscribed about the ellipse $x^2/a^2 + y^2/b^2 = 1$, there are infinitely many with minimum area. Given any point P on the ellipse there is an n-gon of minimum area having P as a point of tangency of one of its sides. If P has coordinates $(a \cos \theta,\ b \sin \theta)$ then the other points of tangency belonging to the other sides of the n-gon are*

$$(a \cos(\theta + 2k\pi/n), b \sin(\theta + 2k\pi/n)) \tag{1}$$

where $k = 1, 2, 3, \ldots, n - 1$.

7.5. Tangent Lines and Extreme Values. We start with a question that is closely related to tangent lines, although not apparently so.

Example 1. Given any constants c and d, find the largest and smallest values of $cx + dy$ taken over all points of the ellipse $x^2/a^2 + y^2/b^2 = 1$.

Solution. Using the parametric equations of the ellipse, $x = a \cos \theta$ and $y = b \sin \theta$ from §7.2, we see that the problem amounts to finding the maximum and minimum of $ac \cos \theta + bd \sin \theta$ among all possible values of θ. In §5.5 it was proved that the extreme values of $A \sin \theta + B \cos \theta$ are $\sqrt{A^2 + B^2}$ and $-\sqrt{A^2 + B^2}$. Replacing A and B by bd and ac, we conclude that the largest and smallest values of $cx + dy$ on the ellipse are

$$\sqrt{a^2c^2 + b^2d^2} \quad \text{and} \quad -\sqrt{a^2c^2 + b^2d^2}, \tag{1}$$

respectively.

This solution includes the case of a circle; so we can also conclude that the largest and smallest values of $cx + dy$ taken over all points of the circle $x^2 + y^2 = a^2$ are

$$a\sqrt{c^2 + d^2} \quad \text{and} \quad -a\sqrt{c^2 + d^2}, \tag{2}$$

respectively. These values are obtained from (1) above by writing $b = a$, where we are presuming that a is positive.

The method used in Example 1, although very brief, tends to obscure the geometry of the situation. The solution gives us no indication of the locations on the ellipse where the largest and smallest values of $cx + dy$ occur. To get at this question, we approach the matter another way.

Example 2. Find the largest and smallest values of $2x + y$ on the circle $x^2 + y^2 = 25$. At what points on the circle do these values occur?

Solution. By formula (2), the largest and the smallest values of $2x + y$ are $5\sqrt{5}$ and $-5\sqrt{5}$, respectively, obtained by using $a = 5$, $c = 2$, $d = 1$. To answer the question about the points on the circle where the extreme values occur, we proceed as follows. The lines $2x + y = k$ form, for different values of k, a family of parallel lines, so-called "contour" lines. Four of these lines are shown in Figure 7.5a, for the values $k = -5\sqrt{5}$, $k = -5$, $k = 4$, and $k = 5\sqrt{5}$. As k increases, for example, from $k = -5$ to $k = 4$, the line moves to the right. At the point Q in the diagram, an intersection point of the circle $x^2 + y^2 = 25$ and the line $2x + y = 4$, the value of $2x + y$ is obviously 4. It follows that the largest value of $2x + y$ on the circle $x^2 + y^2 = 25$ is the largest value of k for which the line $2x + y = k$ has a point in common with the circle. This is the line farthest to the right in Figure 7.5a, the tangent line. Since we already know that the largest value of $2x + y$ on the circle is $5\sqrt{5}$, we conclude that the equation of this tangent line is $2x + y = 5\sqrt{5}$.

To find the point of tangency we solve the equations $2x + y = 5\sqrt{5}$ and $x^2 + y^2 = 25$ to get the point of intersection. Substituting $y = 5\sqrt{5} - 2x$ into $x^2 + y^2 = 25$ we have

$$x^2 + (5\sqrt{5} - 2x)^2 = 25 \quad \text{or} \quad 5x^2 - 20\sqrt{5}x + 100 = 0. \tag{3}$$

This quadratic equation can be factored in the form $5(x - 2\sqrt{5})^2 = 0$, having the solution $x = 2\sqrt{5}$. This with $2x + y = 5\sqrt{5}$ gives $y = \sqrt{5}$, and hence the largest value of $2x + y$ on the circle $x^2 + y^2 = 25$ occurs at the point $(2\sqrt{5}, \sqrt{5})$. Similarly the smallest value occurs at the point $(-2\sqrt{5}, -\sqrt{5})$.

The geometric considerations in the argument above suggest that we can find tangent lines to circles and ellipses by a variation on the method.

THEOREM 7.5a. *The slope of the tangent line at a point (r, s) on the ellipse $x^2/a^2 + y^2/b^2 = 1$ is $-b^2r/a^2s$, except for the two points $(\pm a, 0)$ at the ends of the major axis where the tangent lines are parallel to the y axis. For these lines, with equations $x = a$ and $x = -a$, slope is not defined. In case $a = b$ the curve is the circle $x^2 + y^2 = a^2$, and the slope of the tangent line at (r, s) is $-r/s$.*

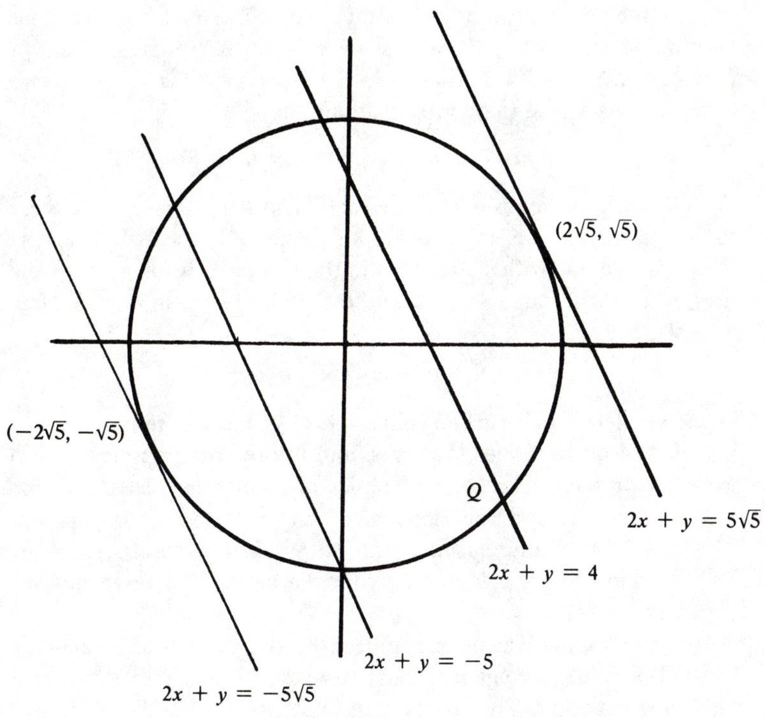

FIG. 7.5a

In the preceding section it is pointed out that the mapping $x = aX$, $y = bY$ carries tangent lines into tangent lines, so we look at the problem in the setting of the circle $X^2 + Y^2 = 1$. The point (r, s) on the ellipse corresponds to the point $(r/a, s/b)$ on the circle, as is readily seen by setting $x = r$, $y = s$ in the equations $x = aX$, $y = bY$, and solving for X and Y. The slope of the tangent line at a point on a circle is easy to calculate because the tangent line is perpendicular to the radius drawn to the point of tangency. So first we get the slope of the line from the center $(0, 0)$ to the point $(r/a, s/b)$, namely, slope sa/rb. The perpendicular line has slope $-rb/sa$, because the product of the slopes of perpendicular lines is -1.

Now to interpret this slope in the xy plane, we need to answer a simple question: If a line has slope m in the XY plane, what is the slope of the corresponding line in the xy plane? To answer this, consider a line $Y = mX + k$, with slope m in the XY plane. The mapping $x = aX$, $y = bY$ is applied in the form $X = x/a$, $Y = y/b$, to give the equation of the correspondling line,

$$y/b = mx/a + k \quad \text{or} \quad y = (mb/a)x + kb.$$

The slope of this line is mb/a, the coefficient of x. Hence the answer to the question above is that a line with slope m in the XY plane corresponds to a line with slope mb/a in the xy plane. It follows that the tangent line with slope $-rb/sa$ in the XY plane corresponds to a line with slope

$$(-rb/sa)(b/a) = -rb^2/sa^2$$

in the xy plane. This is the result claimed in the theorem.

Slopes of tangent lines to curves, and the resulting applications to extremal problems, are central topics of elementary calculus. The method used to find the slope of a tangent line to an ellipse in Theorem 7.5a is quite special, with no application to other curves. This is a topic where calculus is by far the most effective general instrument.

However, before leaving the subject, we draw attention to another method of finding tangent lines that works for any curve having a quadratic equation. The curves with this property are the conic sections: circles, ellipses, parabolas, and hyperbolas. We illustrate the procedure with a hyperbola.

Example 3. Find the slope of the tangent line to the curve $xy = 12$ at the point (4, 3).

Solution. The tangent line at (4, 3) intersects the curve $xy = 12$ at no other point. (This assertion is valid for curves with quadratic equations, but not necessarily for more complicated curves.) The equation of a line with slope m passing through the point (4, 3) is

$$y - 3 = m(x - 4) \quad \text{or} \quad y = mx - 4m + 3.$$

Can we find the value of m such that this line intersects the curve $xy = 12$ only at the point (4, 3)? To find this value, we solve the equation of the line and the curve by substituting $mx - 4m + 3$ for y in $xy = 12$ to get

$$x(mx - 4m + 3) = 12 \quad \text{or} \quad mx^2 + (3 - 4m)x - 12 = 0. \tag{4}$$

Viewing this as a quadratic equation in x, it has two solutions in general. We know that $x = 4$ is a solution for every value of m. Is there a value of m for which this equation has the solution $x = 4$ and no other? This would occur if the quadratic equation has the repeated root $x = 4$, two equal roots that is.

From the theory of quadratic equations we know that $ax^2 + bx + c = 0$ has two equal roots iff $b^2 - 4ac = 0$. Applying this criterion to equation (4) we get

$$(3 - 4m)^2 + 48m = 0, \quad \text{or} \quad 16m^2 + 24m + 9 = 0.$$

The solution of this equation is $m = -3/4$, and this is the required slope.

G5. Use the method of Example 3 to find the slope of the tangent line to the curve $x^2 - 4y - 28 = 0$ at the point $(4, -3)$.

7.6. Shortest Distance from a Point to a Curve. It is suggested intuitively by Figure 7.6a that, if Q is the point on a curve closest to a point P, then the line segment PQ is perpendicular to the tangent line to the curve at Q. Of course, the shortest distance from a point P to a curve C is the radius of smallest circle with center P having a point in common with C. This circle touches C, and they share a common tangent at their point of contact Q.

Similarly, if R is a point on a curve such that the distance PR is a maximum, the line segment PR is perpendicular to the tangent to the curve at R. This property applies only to curves contained in a finite region of the plane, as in Figure 7.6a. Furthermore, we are discussing only curves that turn continuously, so that there is a tangent line at each point. Figure 7.6a also shows that a line segment PS that is perpendicular to the tangent at a point S on a curve may not give either the minimum or the maximum distance from P to the curve.

A line such as PQ in Figure 7.6a that is perpendicular to a tangent line at the point of contact is said to be a *normal*, or a *normal line*, to the curve at the point Q. There is a normal to a curve at every point on the curve where there is a tangent line; the two lines are perpendicular.

Although the properties above can be used to find minimum and maximum distances from a point to a curve, we regard their use as a

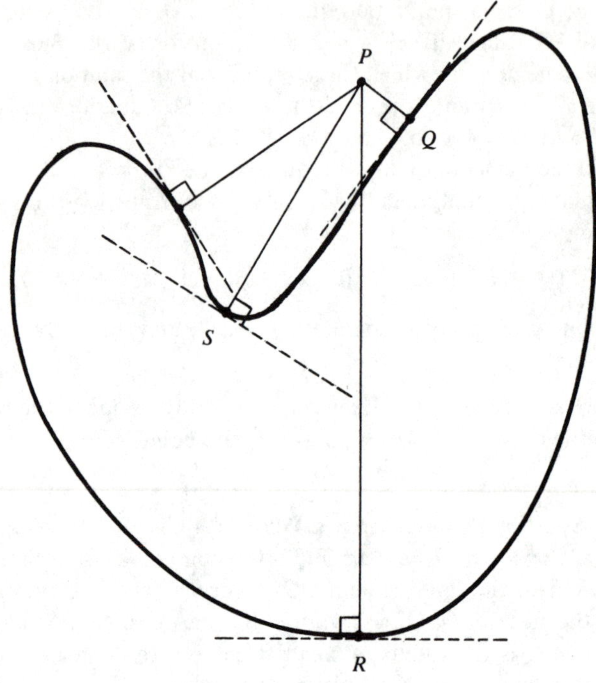

FIG. 7.6a

secondary procedure because we have no general theory of tangent lines. Our primary procedure will be to use the algebraic methods of Chapter 2. This is done in two problems involving shortest distances in §3.4, specifically Problem 4 in the text in that section, and Problem C25. These two problems, like those in the present section, are special cases that can be managed by our limited methods.

The converse of the properties above about minimum and maximum distances does not hold. That is to say, if a line segment PQ is perpendicular to the tangent line at a point Q on a curve, it does not follow that PQ is either the shortest or the longest line from P to the curve (see Figure 7.6a). The second solution of the following example also illustrates this fact.

Example 1. Find the coordinates of all points on the curve $x^2/100 + y^2/25 = 1$ nearest to, and farthest from, the point (0, 9).

Solution. We solve this problem by two methods. The first is an algebraic procedure as in Chapter 2. The square of the distance from (0, 9) to any point (x, y) on the curve is

$$\begin{aligned}(x-0)^2+(y-9)^2 &= x^2+y^2-18y+81\\ &= (100-4y^2)+y^2-18y+81\\ &= 181-3(y^2+6y).\end{aligned}$$

The minimum value of y^2+6y, and hence the maximum value of the distance, occurs if $y=-3$. There are two points on the ellipse with $y=-3$, namely, $(8, -3)$ and $(-8, -3)$. These are the two points on the curve farthest from (0, 9).

To get the point or points on the curve nearest to (0, 9) we want to maximize y^2+6y. This by itself has no maximum, but the problem is to maximize y^2+6y among points on the ellipse, so that y must satisfy the restriction $-5 \leq y \leq 5$. The maximum clearly occurs at $y=5$, and so we get the point (0, 5) on the curve. Thus (0, 5) is the nearest point on the curve, a result that could be obtained from a simple diagram.

As a second method of solving this problem we look for points (x, y) on the curve such that the tangent line at (x, y) is perpendicular to the line segment from (0, 9) to (x, y). By Theorem 7.5a the slope of the tangent line at a point (x, y) on $x^2/100+y^2/25=1$ is $-25x/100y$ or $-x/4y$. The slope of the line from (0, 9) to (x, y) is $(y-9)/x$. Perpendicular lines imply that the product of the slopes is -1 (except for lines parallel to the axes, which we look at in a moment) so that

$$-\frac{x}{4y}\cdot\frac{y-9}{x}=-1,\quad \frac{y-9}{4y}=1,\quad 3y=-9,\quad y=-3.$$

Thus we get the two points $(8, -3)$ and $(-8, -3)$.

But also we must look at cases of lines parallel to the axes. Since (x, y) is on the ellipse, the line from (0, 9) to (x, y) is not parallel to the x axis for any values of x and y, but this line is the y axis itself in case (x, y) is (0, 5) or $(0, -5)$, the ends of the minor axis of the ellipse. Thus there are four normal lines from (0, 9) to the ellipse $x^2/100+y^2/25=1$, the points of tangency being $(8, -3)$,

$(-8, -3)$, $(0, 5)$, and $(0, -5)$. The distances from $(0, 9)$ to each of these are $\sqrt{208}$, $\sqrt{208}$, 4, and 14, respectively. The largest of these is $\sqrt{208}$ and the smallest is 4. Thus the closest point to $(0, 9)$ is $(0, 5)$ and the farthest points are $(8, -3)$ and $(-8, -3)$.

The point $(0, -5)$ on the curve is at neither a maximum nor a minimum distance from $(0, 9)$, and yet the point $(0, 9)$, lies on the normal to the curve at $(0, -5)$.

Some of the problems below require a little care with their interpretation.

G6. Find the points on the ellipse $x^2 + 2y^2 = 2$ that are nearest to, and farthest from, the point $(0, 3)$.

G7. Determine the point or points on the curve $x^2/25 + y^2/9 = 1$ closest to $(k, 0)$, where k is a positive constant.

G8. Which point (or points) on the curve $xy^2 = 54$ is nearest to the origin?

G9. Find the point or points on the hyperbola $2x^2 - y^2 = 2$ nearest to the point $(3, 0)$; nearest to the point $(6, 0)$.

G10. Determine with proof the two points farthest apart on the ellipse $x^2/a^2 + y^2/b^2 = 1$, where as usual we presume that $a > b > 0$. Remark: If the reader feels that this is intuitively obvious, he or she should look at the next problem. More generally, consider the curve

$$\left|\frac{x}{a}\right|^r + \left|\frac{y}{b}\right|^r = 1$$

for $r \geq 2$, which is an ellipse for $r = 2$ and becomes more bulbous in the vicinity of the points $(a, 0)$ and $(-a, 0)$ as r increases. When r gets very large the graph of the equation gets close to the rectangle bounded by the four points $(\pm a, \pm b)$. The maximum distance between two points on the curve is $2a$ in case $r = 2$, but exceeds $2a$ for all real numbers $r > 2$. This last statement can be proved with relative ease by calculus, using Lagrange multipliers.

G11. Find the maximum distance between two points on the curve $x^4/a^4 + y^4/b^4 = 1$.

7.7. Extreme Points on an Ellipse. Consider the problem of finding the highest and lowest points on the ellipse

$$2x^2 + 2xy + y^2 + 4x + 4y - 14 = 0, \tag{1}$$

that is, the point on the ellipse with the largest possible y coordinate and the point with the smallest possible y coordinate. To solve this, we can formulate the equation as a quadratic in x,

$$2x^2 + 2x(y + 2) + y^2 + 4y - 14 = 0.$$

Solving for x in terms of y by using the well-known quadratic formula from elementary algebra, we get

$$x = \frac{1}{2}\left[-y - 2 \pm \sqrt{-y^2 - 4y + 32}\right]. \tag{2}$$

Taken jointly, these two equations (one with the plus sign, one with the minus) are algebraically equivalent to (1) in the following sense. Any point whose coordinates are in the solution set of equation (1) is in the solution set of at least one of the equations (2), and conversely.

It follows that we can use equations (2) to calculate the coordinates of points lying on the ellipse by assigning numerical values to y and solving for x. For example, if we set $y = 0$ in (2) we get $x = -1 \pm \sqrt{8}$. But we can use only those values of y for which $-y^2 - 4y + 32$ is positive or zero. Observing that

$$-y^2 - 4y + 32 = (8 + y)(4 - y)$$

we see that the admissible values of y satisfy the inequalities $-8 \leq y \leq 4$. Hence the largest possible value of y is $y = 4$, and the smallest is $y = -8$. Substituting these into (2) we get $x = -3$ and $x = 3$, respectively. Thus the point on the ellipse (1) with the greatest y coordinate is $(-3, 4)$, and the point with the least y coordinate is $(3, -8)$.

The method developed above can easily be used to derive formulas for the coordinates of the points with largest and smallest y coordinates on an arbitrary ellipse with equation

$$F(x, y) = ax^2 + 2bxy + cy^2 + 2dx + 2ey + k = 0. \tag{3}$$

Although such formulas are not difficult to obtain, we do not develop them here. Whether equation (3) represents an ellipse (or a circle as a special case) depends on the values of the coefficients. It is necessary that

$$a > 0, \quad c > 0, \quad ac - b^2 > 0, \tag{4}$$

but these conditions are not sufficient to guarantee that (3) represents an ellipse. It is also necessary that the constant k is not too large. Again we do not develop any general formulas, although this is easily done. The following example serves to illustrate the procedure.

Example 1. For what values of the constant k does the equation

$$F(x, y) = x^2 - 2xy + 2y^2 - 10x + 4y + k = 0 \tag{5}$$

have (i) infinitely many solutions (x, y) in real numbers, so that the graph of equation (5) is an ellipse; (ii) exactly one solution, so that the graph consists of a single point; (iii) no solutions, so that the graph of equation (5) is the empty set? In case (ii), find the coordinates of the single point.

Solution. Following the procedure developed earlier, we seek the points on the graph of (5) with the greatest and least y coordinates. Solving (5) for x in terms of y we get

$$x = y + 5 \pm \sqrt{-y^2 + 6y + 25 - k}. \tag{6}$$

The expression under the square-root sign can be written in the form

$$-y^2 + 6y - 9 + 34 - k = -(y - 3)^2 + 34 - k. \tag{7}$$

This expression is positive for infinitely many values of y iff $34 - k$ is positive. Thus the answer to part (i) is $k < 34$. In case $k = 34$ the expression (7) reduces to $-(y - 3)^2$, which is negative for all values of y except $y = 3$. Hence the answer to part (ii) of the question is $k = 34$, in which case the equation (5) has a single point (8, 3) as its graph; the value $x = 8$ comes from substituting $y = 3$ into (6).

Finally, it is clear that the equations (6), and so also (5), have no solutions in real numbers if $k > 34$.

G12. Find the coordinates of the point having greatest x coordinate on the ellipse (1) in the text above. Also find the point having least x coordinate.

G13. Find the coordinates of the center of the ellipse $x^2 - 2xy + 2y^2 - 10x + 4y + 9 = 0$, using the information (which is intuitively clear from the graph of an ellipse) that the center is the midpoint of the line segment joining the two points on the ellipse having largest and smallest y coordinates. (Note that the equation given is the special case with $k = 9$ of the equation (5) in the text.)

Notes on Chapter 7

The reader familiar with geometry will recognize the mapping $x = aX$, $y = bY$ as a simple affine transformation. An extensive analysis of the applications of transformations in many areas of mathematics is given in the paper by Klamkin and Newman (1966).

Problem G2, that the sides of a rectangle inscribed in an ellipse are parallel to the axes of the ellipse, is proved in the paper of G. D. Chakerian and L. H. Lange (1971, p. 68). Our proof is similar to theirs. This result is apparently assumed by the authors of some calculus books, who state the problem about the rectangle of largest area inscribed in an ellipse *without* including the proviso that the sides are parallel to the axes. Without this condition, the problem is surely difficult for beginners.

§7.3. For an early paper on polygons of largest area inscribed in an ellipse, somewhat different from our approach, see W. V. Parker and J. E. Pryor (1944).

§7.5. A different approach to tangent lines to $xy = a^2$ has been given by L. H. Lange (1976).

The problem of finding the extreme points on an ellipse, and the solution by a simple algebraic procedure, was suggested by Andy Lau. His interest in this arose from seeking the extreme points of an elliptically-shaped confidence region for two parameters in a regression; cf. N. Draper and H. Smith (1966, pp. 64, 65).

CHAPTER **8**

THE BEES AND THEIR HEXAGONS

8.1. The Two Problems. The structure of a honeycomb in a beehive has long fascinated mankind. It has been discussed with admiration by philosophers such as Aristotle, and measured and analyzed by biologists such as the French naturalist René Reaumur. The outstanding organization necessary for the building of the honeycomb and the gathering of the honey has been celebrated in verse:

> For where's the state beneath the firmament,
> That doth excel the bees for government?

And since upwards of 20,000 trips to the flowers are needed to produce a pound of honey, homage has been paid to the incredible effort involved:

> How doth the busy little bee
> Improve each shining hour.

Mathematicians are fascinated by the geometric structure of the honeycomb, which is composed of a large number of cells whose cross-sections are regular hexagons. These special polygons have the property that they "tile" the plane, as shown in Figure 8.1a. (The overlaid square in this figure can be ignored for the moment; it is needed later.) A polygon is said to *tile*, or *pave*, or *tessellate,* the plane if congruent copies of the figure can be used to cover the plane without any gaps or overlaps.

Pappus of Alexandria had this to say about the possible geometrical configurations that might have been used as alternatives to hexagons in the construction of the hive:

> Now only three rectilineal figures would satisfy the condition, I mean regular figures which are equilateral and equiangular; for the

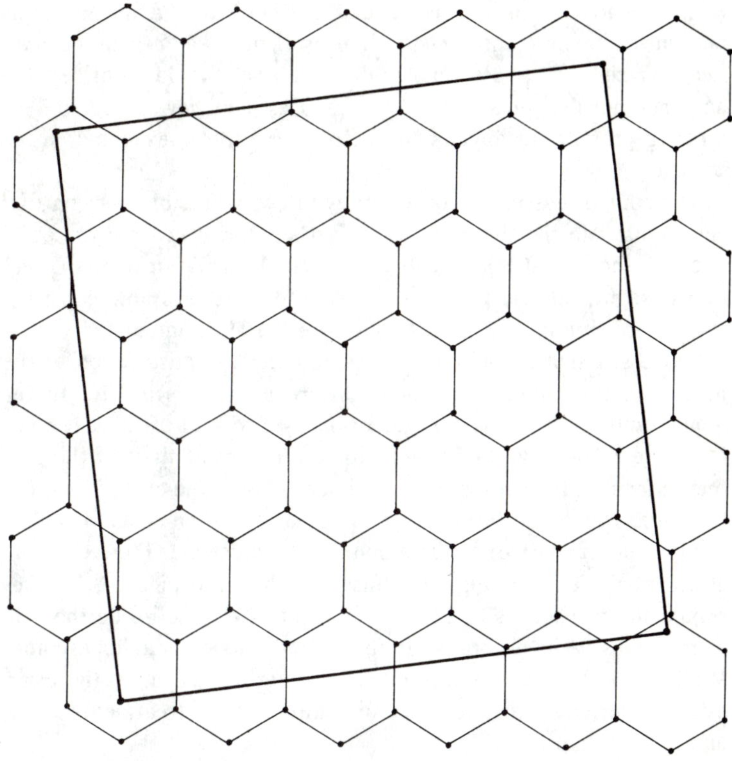

FIG. 8.1a

> bees would have none of the figures which are not uniform .... There being then three figures capable by themselves of exactly filling up the space about the same point, the bees by reason of their instinctive wisdom chose for the construction of the honeycomb the figure which has the most angles, because they conceived that it would contain more honey than either of the two others. Bees, then, know just this fact which is of service to themselves, that the hexagon is greater than the square and the triangle and will hold more honey for the same expenditure of material used in constructing the different figures.

One purpose of the present chapter is to give reasons why there are only three figures that meet the specifications laid down by Pappus, namely, that the polygon must be regular. But going beyond this, we want to consider nonregular polygons also. After all, if the bees have contrived this construction by virtue of a "certain geometrical foresight," as Pappus contended, we should consider the anthropomorphic possibility that the bees somehow eliminated all other geometric arrangements before choosing hexagonal cross-sections.

Thus our discussion is concerned with every possible polygon that can be used to tile the plane, and further, to select from these the one with the largest isoperimetric quotient. It turns out to be the regular hexagon; hence, the bees do use the most favorable construction for maximum storage of honey with the least amount of wax.

Our discussion is limited to the cross-section construction of the hive, with the regular hexagonal pattern as in Figure 8.1a. In the honeycomb, each cell is a 3-dimensional hexagonal prism, as shown in Figure 8.1b. Now the *second* problem in this analysis of the geometric form of the cells is the question of closing the end of each cell with a lid which will cap the given prismatic column so as to enclose the maximum volume for the amount of wax used. If these ends are closed by three congruent rhombuses, as shown in Figure 8.1c, they should meet at angles of 120° for maximum volume inside the cell. This is closely in accord with the actual shape of a honeycomb. While these 3-dimensional problems are not treated here, the interested reader can find references to comprehensive treatments at the end of this chapter.

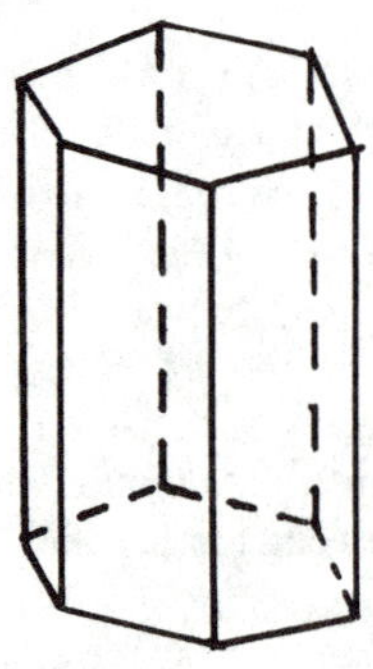

FIG. 8.1b

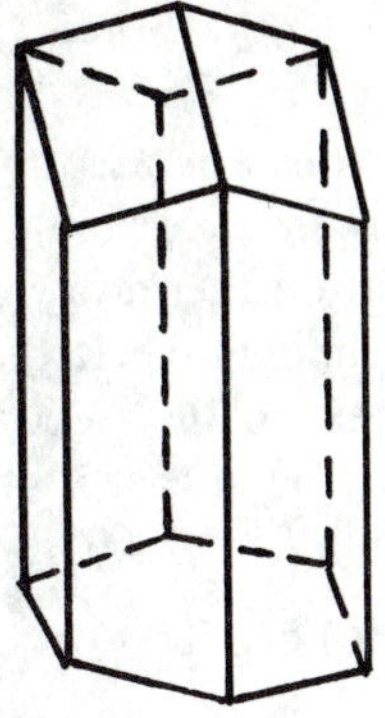

FIG. 8.1c

8.2. Tiling by Regular Polygons. In this and the next two sections the possibilities for tiling the plane by various polygons are examined; regular polygons first, then nonconvex polygons, and finally convex polygons in general. The subject is extensive, and only the high spots are treated here. Moreover, this topic has its unsolved problems, which of course adds to the interest because the story is not yet fully told.

As noted by Pappus, *there are only three kinds of regular polygons that can be used to tile the plane, namely, the equilateral triangle, the square, and the regular hexagon.* It should be kept in mind that we are talking about tilings involving replicas of a single polygon, so that all the polygons used in a tiling are congruent. The fact that regular hexagons can be used to tile the plane is illustrated in Figure 8.1a. Similarly, it is easy to see that the plane can be covered by squares, and also by equilateral triangles.

The fact that no other regular polygon will do is easily demonstrated by a consideration of interior angles. Let the vertices of a regular n-gon be $P_1, P_2, \ldots, P_n$. Extend the side P_1P_2 a little beyond P_2, and similarly the side P_2P_3 a little beyond P_3, and so on around the figure, creating n angles just outside the polygon. Each of these exterior angles has size $180° - \theta$, where θ denotes the size of an interior angle. These external angles have a sum of $360°$, and hence we obtain

$$n(180° - \theta) = 360°, \qquad \theta = 180° - \frac{360°}{n}. \tag{1}$$

Now if the regular n-gon can be used to tile the plane, the angles must fit together precisely so that there are no overlaps or gaps. If exactly k of the n-gons fit together at a vertex, then it must be that $k\theta = 360°$. For example, in Figure 8.1a there are three hexagons fitting together at each vertex; so $3\theta = 360°$ or $\theta = 120°$, and this is the value of θ in equations (1) in case $n = 6$.

Substituting θ from (1) in $k\theta = 360°$, we get

$$k(180° - 360°/n) = 360° \quad \text{or} \quad k(1 - 2/n) = 2$$

by dividing by 180°. This equation can be written in various forms, such as

$$k(n-2) = 2n, \quad (k-2)(n-2) = 4.$$

Since $n \geq 3$, this equation shows that $n - 2$ is a positive divisor of 4. The only positive divisors of 4 are 1, 2, and 4, giving

$$n - 2 = 1, \quad n - 2 = 2, \quad \text{or} \quad n - 2 = 4$$

as the only possibilities. These lead to $n = 3$, $n = 4$, or $n = 6$, as was claimed.

Thus among regular polygons for the construction of a hive, the bees could have used equilateral triangles, squares, or regular hexagons. Among these three the last one has the largest isoperimetric quotient, and in this sense it is the best candidate among regular polygons for the construction of the hive.

8.3. Nonconvex Polygons. Although it is not possible to tile the plane with any convex polygon having seven or more sides, as proved in the next section, it is a different matter with nonconvex polygons. Figure 8.3a shows a nonconvex octagon *ABCDEFGH* that can be used to tile the plane. The line segments *DE* and *AH* are equal and parallel, and the quadrilaterals *ABCD* and *HGFE* are congruent.

However, if we draw the lines *AD* and *HE*, the rectangle *ADEH* is obtained from the octagon by filling in the indented part *HGFE* with the protruding part *ABCD*. Furthermore, the rectangle *ADEH* has the same area as the octagon but a smaller perimeter. Hence the rectangle *ADEH* has a greater isoperimetric coefficient than the oc-

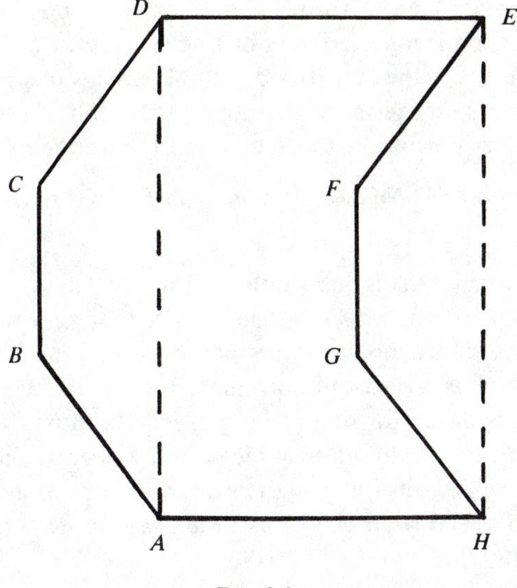

FIG. 8.3a

tagon *ABCDEFGH*. Also we note that any rectangle can be used to tile the plane.

These considerations suggest that in order to establish the desired result that the regular hexagon has the largest isoperimetric quotient among polygons that can be used to tile the plane, it suffices to consider convex polygons only.

8.4. Tiling by Convex Polygons. Any triangle can be used to tile the plane, and also any quadrilateral. The proofs of these results are left to the reader in the problems below. Some convex pentagons, but not all, can be used to tile the plane, and the same is true of convex hexagons. However, whereas the problem of precisely which convex hexagons can be used has been solved, the analogous question for pentagons has not been completely settled.

One example of a hexagon that can be used to tile the plane is readily given, the regular hexagon. An example of a pentagon that tiles the plane can be derived by joining the straight line segment between the midpoints of two opposite sides of a regular hexagon. The hexagon is separated into two congruent pentagons, which can be used to tile the plane. This is just one easy example, of course. The general question is to describe or characterize every convex pentagon that can tile the plane. Further details about this unsolved problem, with references, are given in the notes at the end of the chapter.

Our primary purpose in this section is to prove the following result:

THEOREM 8.4a. *No convex n-gon can be used to tile the plane if* $n \geq 7$.

We prove this result here only in the case of what is called a "strict" tiling, or an "edge-to-edge" tiling, that is, a tiling in which any two tiles with a point in common have (i) exactly one point in common which is a vertex of both tiles, or (ii) a line segment in common which is an entire side of both tiles. The proof in the general case involves essentially no new ideas, but a slightly different interpretation is needed in the way the sides and vertices are counted. A reference to a readily accessible proof is given in the notes at the end of the chapter.

Considering just edge-to-edge tilings then, suppose there is a convex n-gon P that can be used to tile the plane, where n is a fixed integer ≥ 7. Let the perimeter of the polygon P be taken as the unit of length, and let A denote the area of P. The plane having been tiled with replicas of P, let us impose a coordinate system and then define S_r to be the square (including the interior as well as the four sides) whose vertices are $(\pm r, \pm r)$ with all four combinations of signs. Here r is a positive number which will be chosen sufficiently large to give a contradiction.

(Figure 8.1a can be used to visualize the situation, where the square overlaid on the hexagons can be thought of as S_r. The defect in using this figure as a model is of course that it pictures hexagons, whereas the polygon P in the theorem under discussion has more than six sides. On the other hand, if we could draw a model with the polygon P and its congruent replicas having more than six sides, the theorem would not be true. So Figure 8.1a will have to suffice, with a little imagination applied.)

We will also use the squares S_{r+1} and S_{r+2}, again including their interiors, defined by the vertices

$$(\pm(r+1), \pm(r+1)) \quad \text{and} \quad (\pm(r+2), \pm(r+2)),$$

respectively, as illustrated in Figure 8.4a. Let N_1 denote the union of polygons, including their interiors, having at least one point in common with S_r, and N_2 the polygons having at least one point in common with S_{r+1}. In Figure 8.4b the two vertical line segments through the points $(r, 0)$ and $(r + 1, 0)$ are parts of the boundaries of the squares S_r and S_{r+1}. Three polygons of a presumed tiling are shown. All three belong to N_2, but only the one on the left belongs to N_1. The distance between the two vertical lines is 1, and this is also the perimeter of each of the three polygons. The greatest distance between any two points on one polygon is $< 1/2$, because the perime-

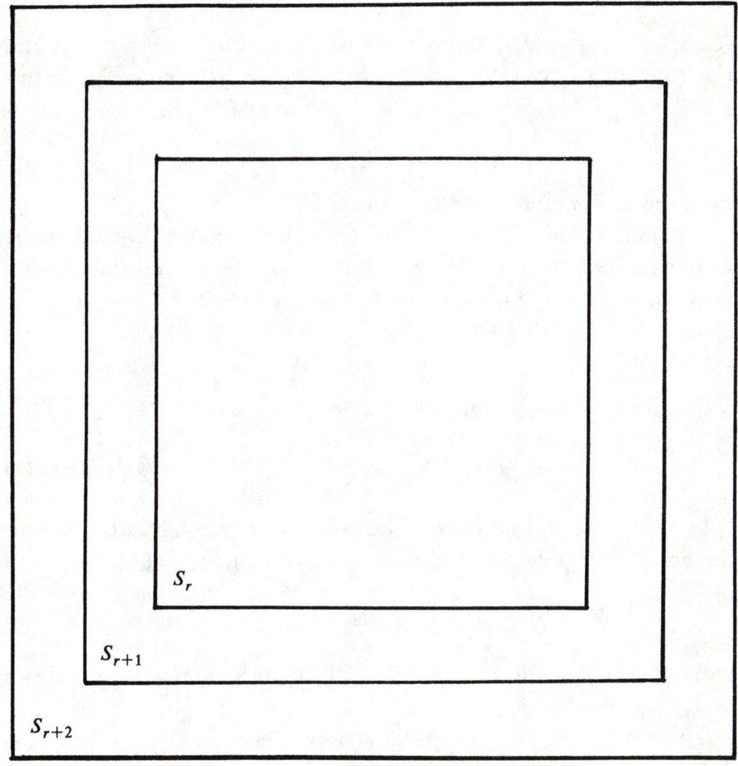

FIG. 8.4a

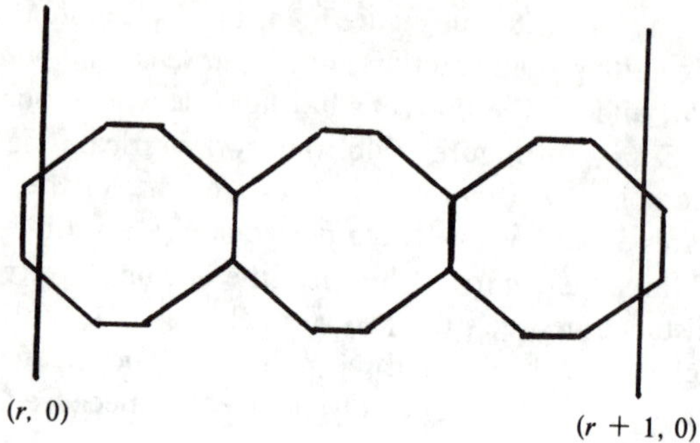

FIG. 8.4b

ter is 1. It follows that all polygons in N_1 lie entirely inside S_{r+1} and don't completely cover it, and similarly all polygons in N_2 lie entirely inside S_{r+2}. This gives a chain of inclusion relationships

$$S_{r+2} \supset N_2 \supseteq S_{r+1} \supset N_1 \supseteq S_r, \tag{1}$$

each region containing the next one.

We now compare the areas of these five regions. Denoting the number of polygons P in the network N_1 by $|N_1|$, and the number in N_2 by $|N_2|$, the areas of N_1 and N_2 are seen to be

$$|N_1|A \quad \text{and} \quad |N_2|A,$$

respectively. Since the area of S_r is $4r^2$, the relations (1) imply

$$4(r+2)^2 > |N_2|A \geq 4(r+1)^2 > |N_1|A \geq 4r^2. \tag{2}$$

Let v denote the number of vertices, each counted only once, in the network of polygons N_1. We prove two more inequalities,

$$2v \geq (n-2)\,|N_1| \quad \text{and} \quad n\,|N_2| \geq 3v, \tag{3}$$

and we show that these contradict (2), thereby completing the proof of Theorem 8.4a.

To show that (2) and (3) yield a contradiction, we multiply the inequalities (3) by $3A$ and $2A$, respectively, and eliminate $6Av$ to get

$$2nA\,|N_2| \geq 3(n-2)A\,|N_1|.$$

Using this and appropriate parts of (2) we then have

$$8n(r+2)^2 > 2nA\,|N_2| \geq 3(n-2)A\,|N_1| \geq 12(n-2)r^2.$$

Ignoring the two middle terms in this chain of four, we divide by 4, expand $(r+2)^2$, and regroup terms to get

$$8nr + 8n > (n-6)r^2.$$

Now n is a fixed integer exceeding 6, so this inequality is false if r is sufficiently large, for example, $r = 8n + 1$.

To complete the proof we establish the inequalities (3). First we calculate the total sum of all the interior angles, in radian measure, of the polygons in N_1. The sum of the interior angles of the n-gon P is $(n-2)\pi$, implying a total sum $(n-2)\pi\,|N_1|$. On the other hand, at each of the v vertices of the network N_1 the total angle is 2π, giving $2v\pi$ as an approximation to the total sum. This approximation is too large, because a vertex on the outside edge of the union N_1 does not contribute a full 2π to the sum of the *interior* angles. It follows that $2v\pi \geq (n-2)\pi\,|N_1|$, giving the first part of (3).

Finally, to prove that $n\,|N_2| \geq 3v$, we make estimates of the number of vertices in the network N_2. Each of the $|N_2|$ polygons in this network has n vertices, and so $n\,|N_2|$ is a multiple count of the vertices in N_2. It is a multiple count because any vertex belonging to more than one polygon in the network is counted more than once. In particular, each of the v vertices of the polygons in the network N_1 is counted at least 3 times for the following reasons. Each polygon in N_1 is an interior polygon of N_2, because N_1 is contained strictly inside S_{r+1}, which in turn lies inside N_2, by (2). Furthermore, each vertex of an interior polygon in N_2 is attached to at least three polygons in N_2, because the tiling polygon P is convex by assumption. It follows that $n\,|N_2| \geq 3v$, thus completing the proof of (3).

H1. Prove that any triangle can be used to tile the plane.

H2. Prove that the plane can be tiled by any quadrilateral, whether convex or not.

8.5. The Summing Up. The arguments in the preceding three sections support the proposition that, *among all polygons that tile* *the plane, the regular hexagon has the maximum* IQ, *that is, has the largest area for a given perimeter.* Theorem 8.4a shows that there is no competition from convex polygons with 7 or more sides. In §8.3 a nonconvex polygon that tiles the plane is related to a convex tiling polygon with a larger IQ. (The argument is by use of an example, but it can be generalized.) There remain the polygons with 3, 4, 5, and 6 sides. In §3.2 and §3.3 it was established that the equilateral triangle and the square have maximum IQ's among 3- and 4-sided figures. An analogous result for hexagons was given in the Theorem 3.5a. Also in Theorem 6.2a it was proved that among regular n-gons the IQ increases with n.

Finally, what about the convex pentagons? In Theorem 4.2b it was proved that the regular pentagon has the largest IQ among 5-sided figures. But this involved the assumption, removed in Chapter 12, that there exists a pentagon with largest IQ. To avoid a forward reference to Chapter 12, it is established in the problem below that corresponding to any convex pentagon there is a convex hexagon with a higher IQ. This result serves to complete the argument without reference to any later chapter.

H3. Given any convex pentagon, prove that there is a convex hexagon with larger IQ. Suggestion: Create a hexagon by cutting off a small isosceles triangle at any vertex of the pentagon. *Remark*: The argument establishes the more general proposition that given any convex polygon there is another polygon having one more side and a larger IQ.

Notes on Chapter 8

§8.1. The quotation on the organization of the bees is from Guillaume de Saluste, 1544–1590. The tribute to the industry of the bees was written by Isaac Watts, 1674–1748, in the work, "Against Idleness and Mischief." Watts was later parodied by the mathematician Charles L. Dodgson, pen name Lewis Carroll, in *Alice's Adventures in Wonderland*, "How doth the little crocodile/Improve his shining tail . . ."

The quotation from Pappus is from Heath (1921, page 390) with the kind permission of Oxford University Press.

A very thorough analysis of the isoperimetric questions involved in the cell structure of the honeycomb is given in the 1964 paper of L. Fejes Tóth with the delightful title, "What the Bees Know and What

They Do Not Know." This article, a write-up of an invited address before the American Mathematical Society, looks into possible closures of the ends of the cells other than the rhombuses used by the bees. See also the paper by Duane DeTemple (1971).

Harry Polachek (1940) gives a historical outline of the problem, including interesting claims and counterclaims about native geometric abilities of the bees.

A brief discussion focusing exclusively on the rhombuses is given by Dörrie (1965, Problem 93). (The last twelve problems in this book are entitled "extremes.")

§8.4 A complete description of all convex hexagons that tile the plane is given in the 1968 paper of R. B. Kershner, and also in the July 1975 article by Martin Gardner. These sources also intended to give a complete analysis of all convex pentagons that tile the plane, but the list turned out to be incomplete, as pointed out by James (1975) in a write-up by Gardner (1975). A recent survey of this situation, including a discussion of the work of Marjorie Rice, has been given by Doris Schattschneider (1978). For a proof of Theorem 8.4a that is not restricted to edge-to-edge tilings, see a paper by Niven (1978).

CHAPTER 9

FURTHER GEOMETRIC RESULTS

9.1. Introduction. This is a continuation of Chapter 3, except that the problems are now generally more difficult. Most of the results are about triangles, and some of them are quite famous, like the problems of Fermat and Fagnano in §9.2 and §9.3. In the last section we discuss the imbedding of a convex region in a rectangle, using a neat argument by Rademacher. This discussion requires a couple of results, not proved here, about convex regions in the plane. Fortunately these results are quite plausible from an intuitive viewpoint.

9.2. A Problem of Fermat. *Fermat's problem, sometimes called Steiner's problem, is to locate the point P in the plane of a triangle ABC so as to minimize the sum of the distances from P to the vertices, i.e.,* $PA + PB + PC$. This can also be called the airport problem: If three cities on a plain are located at A, B, and C, the problem is to find point P to locate a common airport so as to minimize the length of the network of roads leading to the airport from the three cities. First we show how to find the location of P, at least under some circumstances, assuming the existence of a solution. A little information about ellipses is assumed in the following argument.

Suppose that the point P has the property that $PA + PB + PC$ is a minimum, as shown in Figure 9.2a. Holding $PB + PC$ fixed, we look for nearby points such that $PB + PC = QB + QC$. This equation is in effect the definition of an ellipse through P with foci at B and C, this curve being the collection of points such that the sum of the distances from each point on the curve to B and C is a constant. Since P gives a minimum sum, we have

$$PA + PB + PC \leq QA + QB + QC,$$

and it follows that $PA \leq QA$. Thus P is a point on the ellipse such that PA is least, i.e., P is a point on the ellipse closest to A. This implies, as noted in §7.6, that the line PA is perpendicular to the ellipse at P, and so perpendicular to the tangent line HPK to the ellipse at P, as in Figure 9.2a. In this case, then, there is only one such point P, and $\angle APH = \angle APK = 90°$.

Also we will assume the well-known property of the ellipse that the line segments BP and CP from the foci B and C to any point P on the ellipse make equal angles with the tangent line HPK, that is, $\angle BPH = \angle CPK$. (This is the property of the ellipse that is used in the whispering room of science museums so that a person whispering at point B can be heard with ease by a listener at point C, perhaps 30 feet away or so, because the elliptical shape of the room is "reflecting" the sound waves from B to the other focus C.)

From these considerations we have two pairs of equal angles at P, from which $\angle APB = \angle APC$ follows by addition. Thus AB and AC subtend equal angles at P, and by symmetry BC must subtend the same angle again at P. This gives

$$\angle APB = \angle APC = \angle BPC = 120°, \qquad (1)$$

which specifies the location of P to minimize $PA + PB + PC$.

However, to arrive at this conclusion we assumed the existence of a point P for which the sum of the distances $PA + PB + PC$ is a

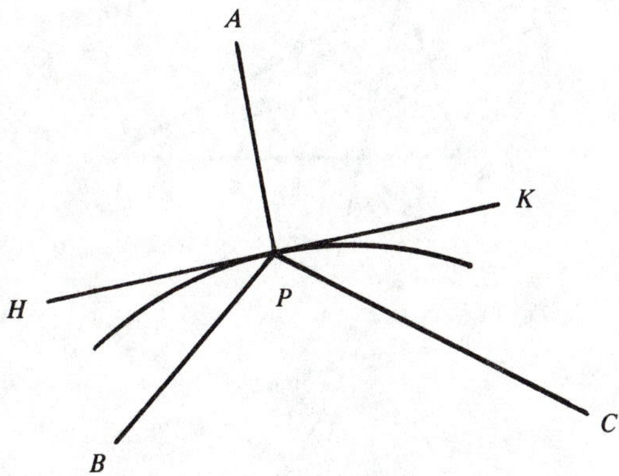

FIG. 9.2a

minimum. Also, a fairly sophisticated property of the ellipse was taken for granted. It turns out, furthermore, that there is not always a point P such that conditions (1) are satisfied. We now state a theorem outlining a more satisfactory solution of the problem and then give a proof that is independent of the preceding analysis. The proof is simpler in terms of the background required.

THEOREM 9.2a. *Given any triangle ABC, there is a unique point P, called the Fermat point of the triangle, such that $PA + PB + PC$ is a minimum. If no angle of the triangle is 120° or larger, the point P is located so that the angles subtended at P by the sides of the triangle are equal, as in equation* (1) *above.* (*An alternative description of the location of P is that it is the common intersection point of the three line segments AA', BB', CC', where the points A', B', C' are situated so that each of the three triangles BCA', CAB', ABC' is equilateral, as shown in Figure* 9.2b.) *On the other hand, if one angle of the triangle ABC is 120° or larger, say $\angle B$, then the point P should be located at B to minimize the sum $PA + PB + PC$.*

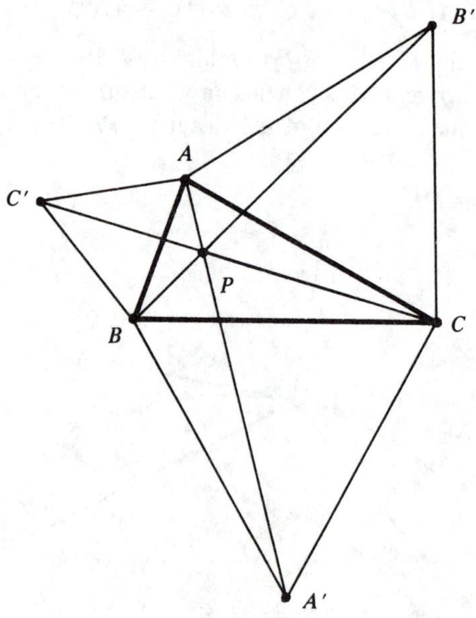

FIG. 9.2b

Part 1. The first part of the proof, where the angles of the given triangle ABC are less than 120°, is rather easy if we use the following simple construction. Let P be any point in the plane of the triangle as shown in Figure 9.2c. With center of rotation B, rotate the line segment BA counterclockwise through 60° to get the position BC', and likewise rotate BP to get BP'. Thus the triangles ABC' and BPP' are equilateral. (It is presumed in this rotation that the vertices A, B, C are in counterclockwise order.)

Since the construction amounts to a rotation of the triangle ABP through 60° about the point B into the position $C'BP'$, we see that $AP = C'P'$. Also $BP = PP'$, and we have

$$AP + BP + CP = C'P' + P'P + PC. \tag{2}$$

So the problem of minimizing the sum on the left side of (2) amounts to minimizing the right side, by choosing the proper location for P. Now C', determined only by the triangle ABC, is independent of P. The sum on the right side of (2) denotes a path from C' to C via the intermediate points P' and P. If we can make this path a straight line, then the right side of (2) is minimized as the length $C'C$. This can be done if P can be located so that each of the angles APB, APC, BPC is 120°. In this case we have

$$\angle C'P'B = \angle APB = 120°, \quad \angle C'P'B + \angle BP'P = 180°$$
$$= \angle BPP' + \angle BPC,$$

and hence $C'P'PC$ is a straight line.

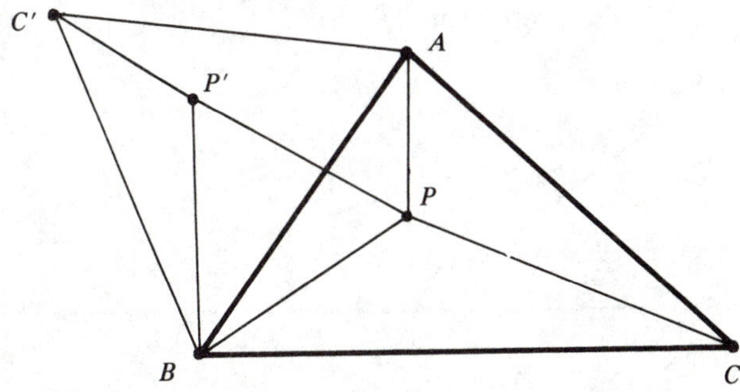

FIG. 9.2c

Finally, we point out that P *can* be located so that the angles subtended at P by the sides AB, AC, and BC are 120°. Consider the circumscribed circles of the equilateral triangles $C'AB$ and $A'BC$: these circles intersect at B of course, but also at another point, and this is the point P which minimizes $PA + PB + PC$. Since the four points C', A, P, B lie on a circle, opposite angles of this quadrilateral add to 180°, giving

$$\angle AC'B + \angle APB = 180°.$$

But $\angle AC'B = 60°$; so we have $\angle APB = 120°$, as wanted. Similarly the points A', B, P, C lie on a circle by the construction above, and we have $\angle BPC = 120°$. It follows that $\angle APC = 120°$, since the total angle at P is 360°.

This optimal location of P is shown in Figure 9.2b, where P is located as explained in the preceding paragraph to satisfy the conditions (1). Now we explain why it is that the location of P can be found in the alternative way described in the theorem. The reason is that just as P lies on the line segment $C'C$, so also it lies on $A'A$ and $B'B$, and hence P is the common intersection of these segments.

Part 2. In case one of the angles of the triangle ABC, say the angle at B, is 120° or larger, the above analysis is not satisfactory. In case $\angle B = 120°$, for example, the points C', B and C are collinear. This suggests that the optimal location for P is at B. In case $\angle B >$ 120°, the construction used above does not lead to a solution of the problem.

So we make a fresh start with a different, but similar, construction. Let P be any point, other than B itself, in the plane of the triangle ABC. We prove that $AB + BC < AP + BP + CP$, as

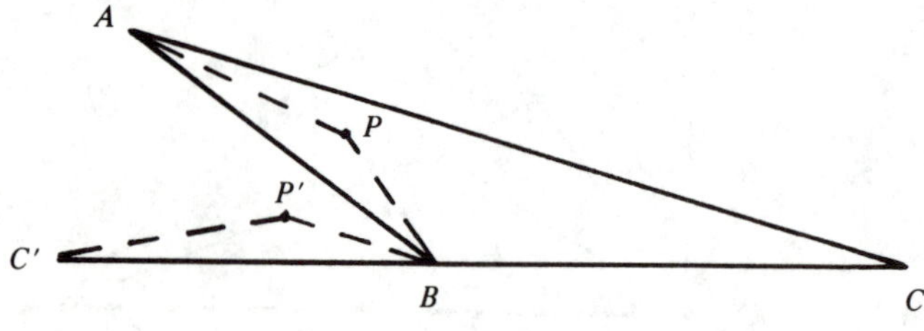

Fig. 9.2d

claimed in the statement of the theorem. As shown in Figure 9.2d, the point C' is chosen on CB extended so that $C'B = AB$. The point P' is chosen so that triangle $C'P'B$ is obtained by rotating the triangle APB through the angle ABC'. (This rotation is counterclockwise if A, B, C are listed in counterclockwise order in the given triangle ABC.) Thus we see that

$$C'P' = AP, \qquad \angle PBP' = \angle ABC' \leq 60°.$$

It follows that

$$BP \geq P'P \quad \text{and} \quad AP + BP + CP \geq C'P' + P'P + PC$$
$$> C'C = AB + BC,$$

since not both P and P' can lie on the line segment $C'C$ by the construction.

An entirely different approach to Fermat's problem is given in §10.4.

Most of the results of the French mathematician Pierre de Fermat (1601–1665) are known only through his correspondence. He stated his theorems in letters to his contemporaries, in most cases without any proofs. It is curious that, although he was perhaps the greatest mathematician of the seventeenth century, he was a lawyer by profession; mathematics was his avocation. He is most famous for the assertion known as "Fermat's last theorem," which states that there are no solutions in positive integers x, y, z of the equation $x^n + y^n = z^n$ for any integer $n > 2$. Whether this is true is not known, although it has been verified for all n from 3 to 100,000.

I1. Given any triangle ABC, locate the point P inside or on the boundary of the triangle so as to maximize the perimeter of the triangle PBC.

I2. Given a convex quadrilateral, where should the point P be located so that the sum of the distances from P to the vertices of the quadrilateral is a minimum?

I3. Solve the preceding problem in the case of a nonconvex quadrilateral.

I4. Let K be any point on the side AB of a triangle ABC with angle $C \geq 90°$. Prove that $AB + CK > AC + BC$.

I5. If P is any point in the interior of the side AB of any triangle ABC, prove that $CP < \max(CA, CB)$.

I6. Given any triangle ABC, locate P inside or on the boundary of the triangle so as to maximize the sum of the distances $PA + PB + PC$.

I7. Locate the point P in the plane of a triangle so that the sum of the lengths of the perpendiculars from P to the sides (extended if necessary) is a minimum.

I8. Given a rectangle $ABCD$, how should the points P and Q be located in a network of the type shown in Figure **I8** so that the length of the network $AP + DP + PQ + BQ + CQ$ is a minimum? (It is presumed that the rectangle has the property $AB \geq AD$.)

I9. Let ABC be a triangle with no angle larger than 120°. Let $C'AB$, $B'AC$, $A'BC$ be three equilateral triangles lying outside the ΔABC, as in Figure 9.2b. Prove that $AA' = BB' = CC'$.

9.3. The Inscribed Triangle. Given any triangle ABC, an *inscribed triangle* is one with one vertex on each side of ABC. It is natural to ask about the inscribed triangles with largest and with smallest areas, and with largest and smallest perimeters.

Two of these questions have obvious answers. The inscribed triangle of largest area is the triangle ABC itself, if we allow ABC to be

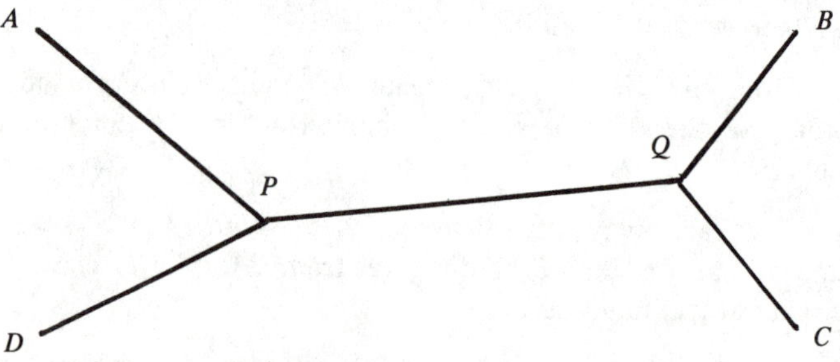

FIG. I8

one of its own inscribed triangles. On the other hand, if we limit inscribed triangles PQR to those having vertices lying in the *interiors* of the sides AB, AC and BC, then there is no inscribed triangle of largest area. Clearly we can get the area of ΔPQR as close to the area of ΔABC as we want, but not equal to it. Similarly, there is no inscribed triangle of smallest area, because although we can get an inscribed triangle PQR with area as close to zero as we want, no triangle can have zero area, by definition.

The question of the inscribed triangle of maximum perimeter is not difficult; it appears among the problems at the end of this section. Finally, we turn to the more difficult question, known as *Fagnano's problem*, of determining the inscribed triangle having minimum perimeter.

THEOREM 9.3a. *Let ABC be an acute-angled triangle, with altitudes AP, BQ, and CR. Then among all triangles inscribed in ΔABC, triangle PQR has the least perimeter. In case one of the angles of ΔABC is 90° or more, there is no inscribed triangle of least perimeter: every inscribed triangle has perimeter exceeding $2h$, where h is the length of the shortest of the three altitudes, and there are inscribed triangles with perimeters as close to $2h$ as we please.*

The triangle PQR described in the theorem, although called the *pedal triangle* of ΔABC by some authors, is more properly called the *orthic triangle* or the *altitude triangle*. The intersection point of the altitudes is called the *orthocenter*. There is a *pedal triangle* of any triangle ABC corresponding to every point X in the plane, determined by the feet of the perpendiculars from X to the sides of ΔABC. Thus the orthic triangle is a special pedal triangle, the one with X located at the orthocenter.

To prove the first part of the theorem, we use the reflection principle. Let PQR be any inscribed triangle of ΔABC, as in Figure 9.3a. Let AP_1Q be the mirror image of ΔAPQ in the line AC, and AP_2R the mirror image of ΔAPR in the line AB. By the congruence of triangles we have

$$PQ + QR + RP = P_1Q + QR + RP_2 \geq P_1P_2, \tag{1}$$

with equality here iff P_1, Q, R, and P_2 are collinear.

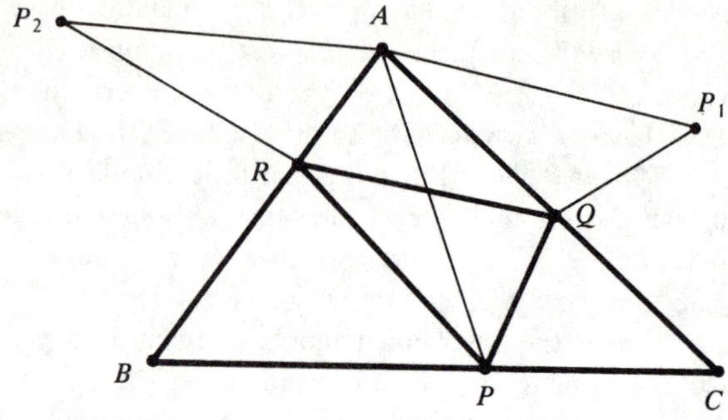

FIG. 9.3a

Note that P_1 and P_2 are determined by P, with no dependence on Q and R. It follows from (1), therefore, that $PQ + QR + RP$ can be minimized if P, Q, and R can be chosen so that the following conditions can be met: (i) that P_1, Q, R, and P_2 are collinear, and (ii) the distance P_1P_2 is a minimum among the infinity of lengths arising from different positions of P. We now prove that these two conditions can be satisfied.

For any given position of P on BC, we choose the points Q and R on the sides AB and AC as the points of intersection of these sides with the line segment P_1P_2. (It should be noted that at this juncture the hypothesis that the triangle ABC has acute angles is being used. The question is, how do we know that the line segment P_1P_2 intersects the segments AB and AC? The answer has two parts. First, from the mirror image triangles we see that

$$\angle P_1AQ = \angle QAP \quad \text{and} \quad \angle PAR = \angle RAP_2. \tag{2}$$

Hence the angle P_1AP_2 is twice the angle BAC, and so $\angle P_1AP_2 < 180°$. Second, because angle C is acute the point P_1 lies on the same side of the line BC as the point A. Similarly, P_2 lies on the same side of BC as A. Thus the line segment P_1P_2 intersects the line segments AB and AC.)

Finally we reach the question, how should P be located on the segment BC so as to minimize P_1P_2? From the congruence of the mirror image triangles, we see that

$$AP_1 = AP = AP_2.$$

If α denotes the angle BAC, then $\angle P_1AP_2 = 2\alpha$ as we have seen, and the law of cosines applied to the triangle P_1AP_2 gives

$$\begin{aligned}(P_1P_2)^2 &= (AP_1)^2 + (AP_2)^2 - 2(AP_1)(AP_2)\cos 2\alpha \\ &= 2(1 - \cos 2\alpha)(AP)^2.\end{aligned}$$

Since α is a constant, to minimize the length P_1P_2 we look for the minimum of AP, and this occurs iff AP is the altitude of the triangle ABC from the vertex A.

Thus, if we locate P so that AP is the altitude from A, the mirror image points P_1 and P_2 give a minimum for the distance P_1P_2. Then Q and R are taken as the intersection points of P_1P_2 with AC and AB, respectively. Relation (1) is an equality with these choices, and we have obtained a *unique* inscribed triangle PQR of least perimeter. Since in the derivation of relation (1) we could have started equally well with B or C playing the role of A, the uniqueness of the triangle PQR guarantees that BQ and CR are the altitudes from B and C. This completes the proof of the first part of the theorem.

To prove the second part, suppose that the angle A is 90° or larger. Starting with the case where angle A is obtuse, we note that the construction of the mirror image points P_1 and P_2 leads to a diagram as in Figure 9.3b. The line segment P_1P_2 does not intersect the segments AB and AC. We prove that

$$PQ + QR + RP = P_1Q + QR + RP_2 > P_1A + AP_2 = 2AP. \quad (3)$$

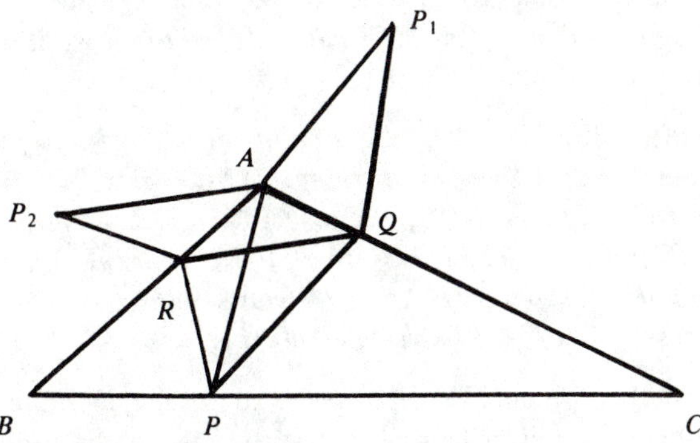

FIG. 9.3b

Inasmuch as the equalities here follow directly from the construction, only the inequality requires an explanation. Consider two circles with centers at P_1 and P_2, having the equal radii P_1A and P_2A. Besides the intersection point A, these circles also intersect at a second point lying on the other side of the line of centers P_1P_2. The sum $P_1Q + QR + RP_2$ is the length of a path from P_1 to P_2 passing outside the two circles, and it follows that this length exceeds the sum $P_1A + AP_2$ of the two radii.

Having proved (3), we note that the least value of AP is the altitude h from A to BC. Thus for all inscribed triangles PQR we have $PQ + QR + RP > 2h$, and this inequality gets as close as we please to an equality by taking P to be the foot of the perpendicular from A to BC, and locating Q and R very close to A. Of course, PQR is not an inscribed triangle if Q and R coincide with A.

We turn finally to the case where $\angle A = 90°$. The construction illustrated in Figures 9.3a and 9.3b must now be modified so that P_1, A, and P_2 are collinear. Hence we see that for any inscribed triangle PQR,

$$PQ + QR + RP = P_1Q + QR + RP_2 > P_1P_2 = 2AP,$$

with a strict inequality here because Q and R cannot coincide with A. Locating the point P so that $AP = h$, we conclude again that $PQ + QR + RP > 2h$.

Having completed the proof of Theorem 9.3a, we establish now a property of any inscribed triangle. Any inscribed triangle PQR separates the parent triangle ABC into four smaller triangles. Among these four triangles, PQR cannot have the least area, except in a very special circumstance.

THEOREM 9.3b. *Let P, Q, R be points in the interior of the sides BC. AC, AB, respectively, of a triangle ABC. Let x, y, z, w be the areas of the triangles ARQ, BRP, CPQ, PQR, respectively, as shown in Figure* 9.3c. *Then* $w > \min(x, y, z)$ *unless P, Q, R are the midpoints of the sides of ΔABC, in which case $x = y = z = w$. (In this case, PQR is called the medial triangle.)*

To prove this, we note first that given any triangle ABC it is possible to define the unit of length so that the area of the triangle is 1.

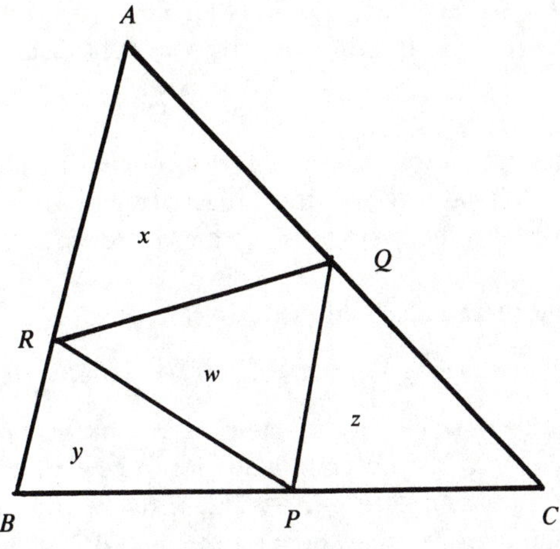

FIG. 9.3c

(This is problem A3 of Chapter 1.) Thus we may assume that $x + y + z + w = 1$. Let p, q, r be the ratios of lengths defined by

$$p = \frac{BP}{BC}, \quad q = \frac{CQ}{CA}, \quad r = \frac{AR}{AB}.$$

This definition of q implies $AQ = (1 - q)AC$, and the area of ΔARQ is

$$x = \frac{1}{2}(AR)(AQ) \sin A = r(1 - q)\frac{1}{2}(AB)(AC) \sin A = r(1 - q).$$

Similar results hold for triangles BPR and CPQ, and so we have

$$\begin{aligned} x &= r(1 - q), \quad y = p(1 - r), \quad z = q(1 - p), \\ w &= 1 - x - y - z = (1 - p)(1 - q)(1 - r) + pqr. \end{aligned} \tag{4}$$

Applying the inequality $a + b \geq 2\sqrt{ab}$ to the last sum, we have

$$w \geq 2[pqr(1 - p)(1 - q)(1 - r)]^{1/2} = 2(xyz)^{1/2}. \tag{5}$$

Now assume that $w \leq \min(x, y, z)$. To prove the theorem we first show that $x = y = z = w = 1/4$. Without loss of generality we may presume that $z = \max(x, y, z, w)$ so that $z \geq 1/4$. Then $x \geq w$ and $y \geq w$ imply $xy \geq w^2$, and this with the square of inequality (5) gives

$$xy \geq w^2 \geq 4xyz \geq xy.$$

It follows that we must have equality throughout, and so in particular $z = 1/4$. That is to say, the largest of the four areas x, y, z, w is 1/4, which is also the average of the areas. Hence $x = y = z = w = 1/4$.

Substituting these values into (5), we have

$$pqr(1-p)(1-q)(1-r) = 1/64. \tag{6}$$

The largest value of $p(1-p)$ among positive numbers $p < 1$ is 1/4, occurring iff $p = 1/2$. The same holds for $q(1-q)$ and $r(1-r)$, and hence (6) implies $p = q = r = 1/2$. It follows that P, Q, and R are the midpoints of the sides of the triangle ABC, as we wanted to show.

I10. Let P, Q, R be any points lying in the interiors of the sides BC, CA, AB, respectively, of a triangle ABC. Is it true for all triangles and all possible locations of the points P, Q, R that the perimeter of ΔABC exceeds the perimeter of ΔPQR? If so, prove it; if not give a counterexample.

I11. Let K be a point in the interior of a triangle ABC. Extend AK, BK, CK to intersect the opposite sides of the triangle at P, Q, R, respectively. Is it true for all triangles and all locations of K that

$$AB + BC + CA > AP + BQ + CR?$$

If so, prove it; if not, give a counterexample.

9.4. A Theorem of Erdös and Mordell.

As a preamble to the main result of this section, we give a famous result of Euler.

THEOREM 9.4a. *If r and R are the radii of the inscribed and circumscribed circles of a triangle, then $R \geq 2r$, with equality if and only if the triangle is equilateral.*

Here is a very simple proof. Let L, M, and N be the midpoints of the sides BC, AC, and AB, respectively, of any triangle ABC. Then the triangle LMN is similar to the triangle ABC, as is well known, and any length on the triangle LMN is exactly half the corresponding length on ΔABC. Since R is the radius of the circumscribed circle of ΔABC, it follows that $R/2$ is the radius of the circumscribed circle of ΔLMN. But the inscribed circle of the triangle ABC is the smallest circle that has a point in common with each of the three sides. Hence $r \le R/2$.

Furthermore, there is equality here iff the circumscribed circle of triangle LMN is tangent to all three sides of the triangle ABC, i.e., is the inscribed circle of ABC. It is not difficult to see that in this case triangle ABC is equilateral.

This proof is easily extended to a 3-dimensional version of Euler's theorem, relating the radii of the inscribed and circumscribed spheres of a tetrahedron. This is given in Chapter 11.

Before leaving Theorem 9.4a, we point out that another way to prove it is by establishing the geometric result

$$CI^2 = R(R - 2r) \tag{1}$$

for any triangle, where CI is the distance from the circumcenter C to the incenter I. The centers C and I of the circumscribed and inscribed circles coincide iff the triangle is equilateral, and hence CI^2 is positive except in this case. Hence $R - 2r$ is positive except in the case of an equilateral triangle. However, this sketch of an alternative proof of Euler's theorem is incomplete, because we do not prove the relation (1) here.

THEOREM 9.4b (Erdös-Mordell). *Let R_1,R_2,R_3 be the distances to the three vertices of a triangle from any interior point P of a triangle. Let r_1,r_2,r_3 be the distances from P to the three sides. Then*

$$R_1 + R_2 + R_3 \ge 2(r_1 + r_2 + r_3), \tag{2}$$

with equality iff the triangle is equilateral and P is its centroid.

Note that the preceding theorem is not a special case of this result.

To prove the theorem let PL, PM, PN be the perpendiculars from P to the sides BC, CA, AB of a triangle ABC, as in Figure 9.4a. Define

$$R_1 = PA, \quad R_2 = PB, \quad R_3 = PC, \quad r_1 = PL, \quad r_2 = PM, \quad r_3 = PN.$$

Let α,β,γ denote the angles at A, B, C, respectively, in the triangle ABC. Applying the law of sines to ΔAMN and to ΔAMP we get

$$(AM)\sin\alpha = (MN)\sin\angle ANM, \qquad R_1 \sin\angle APM = (AM)\sin 90°. \tag{3}$$

Now since the angles ANP and AMP are 90°, the points A, M, P, N lie on a circle, so that the angles ANM and APM are equal. Hence if we multiply equations (3) and cancel common elements we get

$$MN = R_1 \sin\alpha.$$

Also, since A, M, P, N lie on a circle we see that $\cos\angle MPN = -\cos\alpha$, and hence the law of cosines applied to MN in the triangle PMN gives

$$MN^2 = r_2^2 + r_3^2 + 2r_2r_3\cos\alpha. \tag{4}$$

Also $\alpha + \beta + \gamma = 180°$, so that

$$\cos\alpha = -\cos(\beta + \gamma) = \sin\beta\sin\gamma - \cos\beta\cos\gamma.$$

Together with (4) this gives

$$\begin{aligned} MN^2 &= (r_2\sin\gamma + r_3\sin\beta)^2 + (r_2\cos\gamma - r_3\cos\beta)^2 \\ &\geq (r_2\sin\gamma + r_3\sin\beta)^2. \end{aligned} \tag{5}$$

But $MN = R_1\sin\alpha$, so we have

$$R_1 \geq r_2(\sin\gamma/\sin\alpha) + r_3(\sin\beta/\sin\alpha). \tag{6}$$

Adding this to the analogous inequalities for R_2 and R_3, we collect terms in r_1, r_2, r_3 and then apply the inequality $x + 1/x \geq 2$ for any

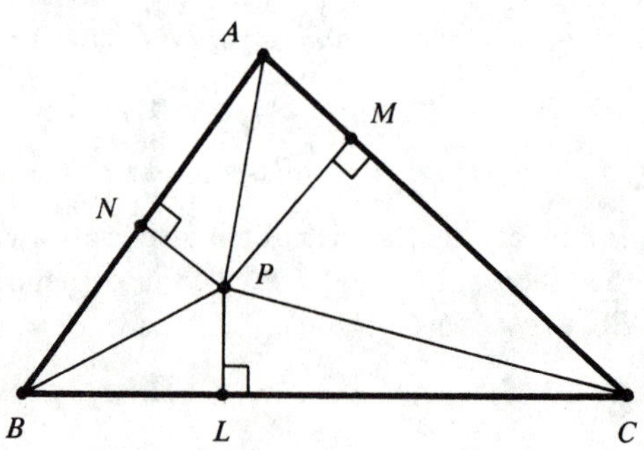

FIG. 9.4a

positive number x, with equality iff $x = 1$. Using this with $x = \sin\gamma/\sin\alpha$, $x = \sin\alpha/\sin\beta$, and $x = \sin\beta/\sin\gamma$, we get the inequality (2) in Theorem 9.4b.

But we must also show that there is equality in (2) iff the triangle is equilateral and P is its centroid. It is easy to see that these conditions imply $R_1 = R_2 = R_3$, $r_1 = r_2 = r_3$, and $R_1 = 2r_1$, giving equality in (2).

Conversely, suppose there is equality in (2). From the argument used above we see that (5) must be an equality, from which it follows that

$$r_2 \cos\gamma = r_3 \cos\beta. \tag{7}$$

Also, from the remark following (6) we see that $\sin\gamma/\sin\alpha = 1$ or $\sin\gamma = \sin\alpha$. Similarly, we get the equations $\sin\alpha = \sin\beta = \sin\gamma$, from which it follows that $\alpha = \beta = \gamma = 60°$, since these are the angles of a triangle. Thus (7) gives $r_2 = r_3$, and similarly we have $r_3 = r_1$ and $r_1 = r_2$. This implies that P is the center of the inscribed circle, which for an equilateral triangle is the same as the centroid.

The results of this section, attributed to Euler, Erdös, and Mordell, comprise only a tiny part of the extensive mathematical work of these three. Leonard Euler (1707–1783) contributed greatly to all branches of mathematics. Many results are named after him, and much of our notation was his creation. For example, he introduced the Greek letter π for the ratio of the circumference to the diameter of a circle, and the letter e for the fundamental constant discussed in §2.8. Paul Erdös and L. J. Mordell, by contrast, are twentieth-century mathematicians, both outstanding problem-solvers as well as theory builders.

9.5. Lines of Division. If P is any point in the interior of a triangle, is it possible to draw a straight line through P that bisects the area? The answer is yes. Moreover, a line can be drawn through P to bisect the perimeter. A line can also be drawn so that P is the midpoint of the segment terminating at the boundary of the triangle. These properties are not peculiar to triangles. The same results hold for any convex polygon, or more generally for any convex region in the plane.

We give a heuristic argument in support of these properties, that is, an argument that is plausible from an intuitive viewpoint. A rigorous mathematical argument depends on a deeper analysis than we wish to consider, involving a study of the real number system and its relationship to lines and curves.

Consider the question raised above, about the straight line through an interior point P that bisects the area of the triangle. The argument in support of this goes as follows. Start with any line through P, intersecting the sides of the triangle LMN at points Q and R, say, as shown in Figure 9.5a. Let A denote the area of the triangle LMN, and let rA be the area of the part of the triangle lying to the left of the line segment QR, more specifically to the left of the segment QR as viewed by an observer at Q looking toward R. In Figure 9.5a, this means that the triangle LQR has area rA, and the area of the quadrilateral $QRNM$ is $(1 - r)A$. If r happens to be $\frac{1}{2}$, then QPR is the line of separation we seek.

If, as is more likely, r is not $\frac{1}{2}$, then we consider what happens as the point Q is moved clockwise around the boundary of the triangle LMN. For example, as illustrated in Figure 9.5a, when Q reaches

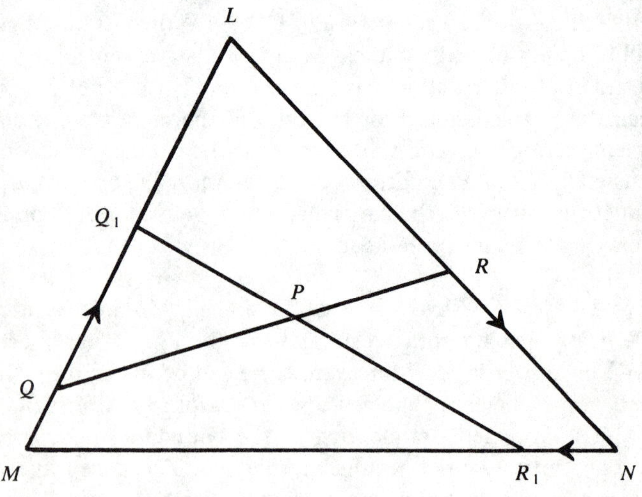

FIG. 9.5a

the position Q_1, then R has moved to R_1, with P fixed, and the points Q_1, P, R_1 collinear. The area to the left of Q_1R_1, that is the area of the quadrilateral Q_1R_1NL in the figure, has the value r_1A, say. Thus as Q moves to position Q_1, the number r changes to r_1. Now consider what happens to r when Q has moved all the way around the corner past L and to the position R along the side LN. The roles of Q and R have been interchanged by this motion, and the area to the left of QR has become $(1 - r)A$.

Thus as Q is moved steadily from its original position around the boundary of the triangle to position R, the area to the left of the moving line segment QR changes from rA to $(1 - r)A$. If r is less than ½, then $1 - r$ exceeds ½, and vice versa. Hence in the continuous change from r to $1 - r$, the value ½ is attained at some intermediate position. At this position the triangle LMN is split into two equal areas.

It is clear that this argument can be applied not simply to a triangular region but to any convex region in the plane, such as a convex polygon. Also, the argument can be used to show that the line segment through the given point P can be chosen so as to bisect the perimeter or, alternatively, so that the segment itself is bisected at P in the sense that $QP = PR$.

The same kind of an argument can be used to prove the following result.

THEOREM 9.5a. *Given any triangle, there is a straight line that bisects both the area and the perimeter. The same result holds for any convex region in the plane, such as a convex polygon.*

Remark. In the case of a triangle there may be more than one straight line that bisects both the area and the perimeter. The details are as follows. Let the lengths of the sides be a, b, c, say with $a \leq b \leq c$. Let s be the semi-perimeter: thus $s = (a + b + c)/2$. Then there are exactly three, two, or one lines that bisect both the area and the perimeter in case $s^2 > 2bc$, $s^2 = 2bc$, or $s^2 < 2bc$, respectively. This result will not be proved here.

As with the earlier argument, the proof we now give of Theorem 9.5a is accurate as far as it goes, but it presumes the so-called intermediate value theorem that a continuous function takes on all values between two given values of the function. Take any point Q on the boundary of the triangle. Let Q' be the point so that Q and Q' bisect the perimeter. If A denotes the area of the triangle, let rA be the area of the portion to the left of the line segment QQ' as viewed by an observer at Q looking toward Q'. If it happens that $r = 1/2$, then QQ' is the line we seek.

In the more likely case that $r \neq 1/2$, let Q be moved steadily counterclockwise along the boundary of the triangle, with Q' moving at exactly the same rate along the boundary, also in the counterclockwise direction. (There is nothing special about the "counterclockwise direction" here; the main thing is to move both Q and Q' in the same direction so that they bisect the perimeter.) When Q has moved halfway around the boundary into the original position of Q' (and so Q' has moved into the original position of Q) the area to the left of QQ' has changed from the original value rA to $A - rA$ or $(1 - r)A$. If $r < 1/2$ then $1 - r > 1/2$, and if $r > 1/2$ then $1 - r < 1/2$. Hence at some intermediate point in the motion the line segment QQ' bisects both the area and the perimeter.

I12. Consider the separation of a triangle into two parts by a line drawn through its centroid C. If r denotes the ratio of the areas of the two parts, prove that the smallest and largest values of r are $4/5$ and $5/4$ for all possible lines of division through C.

I13. In the preceding problem any line segment QCR through the centroid C meets the boundary of the triangle at Q and R. Determine with proof the smallest and the largest values of the ratio $s = QC/CR$ into which the line segment is divided at C.

There is an analogous problem to **I12** and **I13** about the separation of the perimeter of the triangle into two parts by any line of division through the centroid C. Let t denote the ratio of the two parts of the perimeter. Now, whereas the extreme values of r in **I12** and s in **I13** are the same for every triangle, this is not the case with the extreme values of t. For example, in an equilateral triangle the extreme values of t are $4/5$ and $5/4$, and in an isosceles right triangle the extreme values of t are $(6\sqrt{2} - 4)/7$, which is approximately 0.64, and its reciprocal. The greatest lower bound of t among all lines of division through the centroid in all triangles is $1/3$. But $t = 1/3$ cannot be achieved. These results are not difficult to establish by elementary methods.

9.6. Enclosing a Convex Region in a Rectangle. A *convex region R* in the plane is one having the basic property that, if P and Q are any two points in R, the entire line segment PQ is contained in R. Familiar examples are circles, triangles, and ellipses, including their interiors. The points inside a circle, such as the points whose coordinates (x, y) satisfy $x^2 + y^2 < 1$, form a convex region. However, our interest here is in *closed* convex regions, for example $x^2 + y^2 \leq 1$, which contain their boundary points.

The set of points (x, y) with $x \geq 0$ and $y \geq 0$ forms a convex region, even though it is unbounded. A region R is said to be *bounded* if all its points can be enclosed in some circle. With a coordinate system imposed, this condition can be stated in the form that there is a constant k so that all the points of the region are contained in $x^2 + y^2 \leq k$. Any bounded convex region has a well-defined area, as might be expected, although we do not prove it here.

We shall also take it for granted that, given any closed bounded convex region R, there are two points P and Q, not necessarily unique, on the boundary of R, such that the distance PQ is a maximum. (If the region is not closed there may not be any such points. For example there are no such points P and Q in the region defined by $x^2 + y^2 < 1$.) A *support line,* or *line of support,* of a closed bounded convex region R is a straight line that contains at least one point of R, and such that the entire region R lies in the half-plane determined by the line and one side of the line. (With a coordinate system imposed, such a half-plane can be described by $ax + by + c \geq 0$ for some appropriate constants a, b, c.) Given a straight line L in the plane, and any closed bounded convex region, there are precisely two lines of support that are parallel to L (unless the region is an interval on L, in which case L itself is the only line of support parallel to L.) With this preamble, we now state our basic result.

THEOREM 9.6a. *Given any closed bounded convex region R in the plane, with area A, it can be circumscribed by a rectangle whose area is at most $2A$.*

As we shall see, this result is best possible in the sense that it becomes false if the factor "2" in the $2A$ is replaced by any smaller number. In fact, taking the convex region to be a triangle, there is no enclosing rectangle of area less than $2A$. This question is related to Problem 3 in Section 3.4, where the rectangle is to be inscribed in the triangle, not circumscribed about it.

To prove the theorem, we select two points P and Q on the boundary of the region R so that the distance PQ is a maximum. (This maximum distance is called the *diameter* of the region R.) It is clear that the entire region is contained on or between the two lines through P and Q that are perpendicular to the line segment PQ, as illustrated in Figure 9.6a. For if this were not so, there would be a pair of points at greater distance than PQ. Thus these lines perpendicular to the segment PQ are support lines of R. Drawing the two support lines that are parallel to PQ, we obtain for R the circumscribing rectangle $BCDE$, where the points H and K lie on the boundary of the region R and also on the lines BE and CD, respectively. Since the points P, K, Q, H lie on the boundary of the convex region, the entire quadrilateral formed by these four points is contained in R. The area of the quadrilateral $PKQH$ is half the area of the rectangle $BCDE$, because PQ is parallel to BE and DC. If A is the area of the region R, the area of the quadrilateral $PKQH$ is at most A, and the area of the rectangle $BCDE$ is at most $2A$, as we wanted to prove.

There is a special case here that should be mentioned. It is possible that the straight line segment PQ is part of the boundary of the region R. This occurs, for example, if R is a triangle or a semicircle.

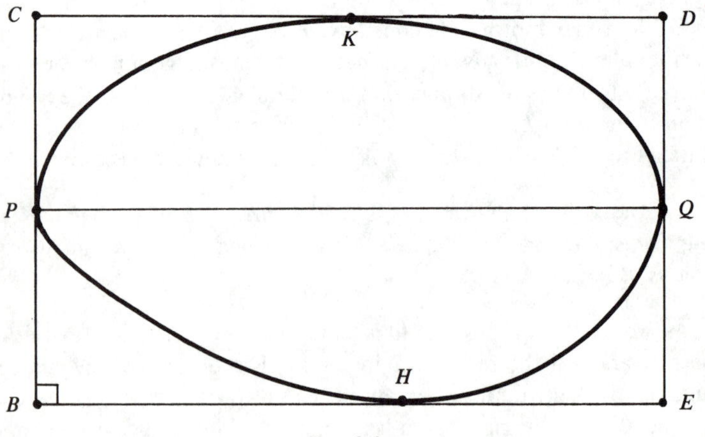

FIG. 9.6a

In this case PQ is itself a support line, and the rectangle $BCDE$ in Figure 9.6a degenerates into $PQDC$ or $PQEB$. The quadrilateral $PKQH$ degenerates into a triangle. However, the final part of the proof of the theorem remains valid.

I14. If the region R in Theorem 9.6a is a triangle, prove that the area of any circumscribing rectangle is at least $2A$, where A is the area of the triangle. Suggestion: It is perhaps easier to prove this result in the equivalent form that, given a rectangle, any triangle formed by three points on its boundary has area at most half that of the rectangle.

I15. Let nine points be given inside or on the boundary of a square of side 1. Prove that three of the points can be selected so that either these points are collinear or the area of the triangle formed is at most $1/8$. (This is a problem from the Chinese Olympiad; see *Amer. Math. Monthly,* 79 (1972) p. 903.)

I16. Given three noncollinear points on the boundary of a parallelogram, what is the largest possible area of the triangle formed compared to the area of the parallelogram?

Notes on Chapter 9

§9.2. The solution of Fermat's problem given here is adapted from Coxeter (1961, pp. 21, 22), who attributes the argument to Hofmann (1929). However, the solution in the case where one angle of the triangle is 120° or more follows the argument given by DeMar (1968) and independently by Sokolowsky (1976). A solution of a different kind by Torricelli can be found in Honsberger (1973). For a brief history of the topic, and a generalization to n points, see Kuhn (1974). Kuhn also discusses the dual problem of the location of a point P to minimize the sum of the distances from P to a given set of lines.

The Fermat problem is only one of a variety of problems, as follows. Let $A_1, A_2, \ldots, A_n$ be distinct points in the plane. The *Fermat problem* is to locate the point P so as to minimize ΣPA_j. The *network problem* is to find connecting lines among the points of minimal total length so that it is possible to move from any point A_i to any other point A_j along the network. The network problem and the Fermat problem are the same for $n = 3$, but not for $n \geq 4$; for a proof see DeMar (1968). The *traveling salesman problem* is to find the path of shortest length starting from A_1, then through each of the other points in any order, and back to A_1. For further details see Dantzig and Eaves (1974) and Gilbert and Pollak (1968).

§9.3. Fagnano's problem, like Fermat's in the preceding section, has an extensive literature. For a discussion of historical background, see Rademacher and Toeplitz (1957, pp. 27-34). For several references to Theorem 9.3b, see O. Bottema et al. (1968, p. 81), and J. Rainwater (1960). The result analogous to Theorem 9.3b for the perimeters of the triangles *ARQ*, *BRP*, *CPQ*, and *PQR* as in Figure 9.3 is also true. For details and proofs of this more difficult result see Bottema et al. (1968, pp. 80-83), and R. Breusch (1962).

§9.4. For the source of the proof given here of Theorem 9.4a, see Fejes Tóth (1948).

The Erdös-Mordell theorem was proposed by Paul Erdös in 1935 and solved by L. J. Mordell in 1937. This has inspired a great deal of subsequent work, such as a neat proof by elementary geometry by Leon Bankoff (1958) and an extension to tetrahedra by N. D. Kazarinoff (1957). For additional results see Bottema et al. (1968, pp. 103–118) and L. Fejes Tóth (1972, pp. 11–14, 190, 191).

§9.5. For a proof of the remark following Theorem 9.5a, see E. P. Starke (1942). Starke's proof involves calculus, and the conditions for one, two, or three lines of separation are given in terms of the angles of the triangle. However, the calculus is not really essential, and the conditions as given here in terms of the relative size of s^2 compared to $2bc$ are included in Starke's proof of the result.

Problem I13 is a special case of the following more general question. Let K be any point in the interior of a convex region R in the plane. Any chord of the region R drawn through K is divided by K into two segments; let ρ be the ratio of the larger segment to the whole chord. There is a maximum such ratio, say $\rho_1(K)$, for each point K in the interior of the region R. Let r be the minimum of $\rho_1(K)$ over all points K in the interior of the region R. Then r is called the *critical ratio* for the closed convex region R, and any point K which divides some chord in the critical ratio is called a critical point. B. H. Neumann (1939) proved that for any closed convex region R in the plane there is a unique critical point and that the critical ratio satisfies the inequality $1/2 \le r \le 2/3$. The value $1/2$ is achieved only for regions with central symmetry (meaning that there is a point K in the region which bisects every chord passing through it) and the value $2/3$ is achieved only for triangles, in which case the centroid is the critical point. A proof that $1/2 \le r \le 2/3$ which generalizes to $1/2 \le r \le n/(n+1)$ for n-dimensional closed convex regions has been given by P. C. Hammer (1951).

§9.6. The proof of Theorem 9.6a, by Hans Rademacher, is adapted from the version given by Pólya and Szegö (1951, p. 109). The corresponding result that every convex region contains a rectangle of at least half its area was proved by Radziszewski (1952). For an analogous result about inscribed parallelograms, see Fulton and Stein (1960). Dowker (1944) has given properties of circumscribed n-gons of minimum area.

CHAPTER **10**

APPLIED AND MISCELLANEOUS PROBLEMS

Various applications of the theory have already been given in earlier chapters, such as the question of tacking a sailboat for maximum speed in §5.6. We now add some further applications, most of which can be studied separately, section by section. However, §10.2 is based on §10.1, and §10.6 on §10.5.

10.1. Lines of Best Fit. Three or more points in the plane do not, in general, lie on a straight line. For example, the points

$$(1, 3), \quad (3, 4), \quad (5, 6), \quad (7, 7) \tag{1}$$

are not collinear. Suppose, however, that these points are expected to lie on a straight line because of an anticipated linear relationship between the coordinates x and y. The problem is to find a straight line that is a good approximation to the set of points, passing very near them if not through them. The method of least squares is one standard way of doing this. The line so obtained is called the *least squares line* or the *linear regression line.*

In this section we describe the method in the case of the given data (1), and then in the next section we derive general formulas to take care of any set of n points. Any straight line not parallel to the y-axis can be written in the form $y = mx + b$, where m is the slope of the line and $(0, b)$ is a point on the line. The problem is to determine the numerical values of m and b so that the line $y = mx + b$ gives a good approximation to the set of data (1).

For the x-values 1, 3, 5, 7, the given data (1) has corresponding y-values 3, 4, 6, 7, respectively. For any line $y = mx + b$ the y-values corresponding to $x = 1, 3, 5, 7$ are

$$m + b, \quad 3m + b, \quad 5m + b, \quad 7m + b,$$

respectively. The method of least squares is a procedure for minimizing

$$(m + b - 3)^2 + (3m + b - 4)^2 + (5m + b - 6)^2 + (7m + b - 7)^2, \tag{2}$$

which is simply the sum of the squares of the differences between the two sets of y-values, those from the given data, and those from the approximating line. In Figure 10.1a the sum (2) is illustrated geometrically as

$$PA^2 + QB^2 + RC^2 + SD^2, \tag{3}$$

where the points A, B, C, D represent the given data (1), and P, Q, R, S are the points on the line $y = mx + b$ with the same x-coordinates as A, B, C, D.

The sum (2) expands into

$$84m^2 + 32mb + 4b^2 - 188m - 40b + 110. \tag{4}$$

Following the method developed in Problems B44, B45, and B46 at the end of Chapter 2, we apply the transformation

$$b = B - 4m \tag{5}$$

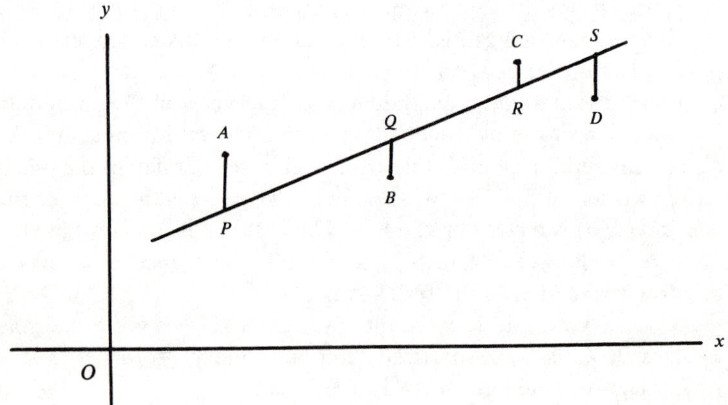

FIG. 10.1a

to reduce the quadratic polynomial (4) to

$$20m^2 + 4B^2 - 28m - 40B + 110. \qquad (6)$$

From Problem B44 we know that $20m^2 - 28m$ is a minimum iff $m = 28/40 = .7$, and $4B^2 - 40B$ is a minimum iff $B = 40/8 = 5$. From $b = B - 4m$ we get $b = 5 - 2.8 = 2.2$. Thus the least square line for the set of data (1) is $y = .7x + 2.2$ or $7x - 10y + 22 = 0$.

10.2. The Least Squares Line in General. In the preceding section the least squares line, or linear regression line as it is called in statistics, was calculated in a specific numerical case. Now we determine the least squares line for a given set of data

$$(x_1, y_1), (x_2, y_2), \ldots, (x_n, y_n) \qquad (1)$$

where it is presumed that the x coordinates are n distinct real numbers. It will be convenient to use the symbols $\bar{x}$ and $\bar{y}$ for the arithmetic means of the x coordinates and the y coordinates, respectively, of the data; thus

$$\bar{x} = \frac{1}{n}\Sigma x_i \quad \text{and} \quad \bar{y} = \frac{1}{n}\Sigma y_i \qquad (2)$$

where the sums here and throughout this section are taken over $i = 1, 2, 3, \ldots, n$. Note that the first of equations (2) can be written in the form

$$\Sigma (x_i - \bar{x}) = 0. \qquad (3)$$

For any straight line $y = mx + b$ the calculated values of y for the x-coordinates of the given data (1) are

$$mx_1 + b, mx_2 + b, \ldots, mx_n + b.$$

Summing the squares of the differences between these calculated values of y and the given data $y_1, y_2, \ldots, y_n$, we get

$$(mx_1 + b - y_1)^2 + (mx_2 + b - y_2)^2 + \cdots + (mx_n + b - y_n)^2. \qquad (4)$$

This sum, which is just the extension of the sum (2) of the preceding section, is to be minimized by appropriate choices of m and b. Thus m and b will be expressed in terms of the values of x_i and y_i from the given data. The outcome is as follows.

THEOREM 10.2a. *The least squares line for the data* (1) *is* $y = mx + b$, *where* m *and* b *are determined by*

$$m = \frac{\Sigma\, y_i(x_i - \bar{x})}{\Sigma\, (x_i - \bar{x})^2} \quad \text{and} \quad b = \bar{y} - m\bar{x}. \tag{5}$$

As an application of this result, we can verify the calculations of the preceding section for the data

$$(1,3), \quad (3,4), \quad (5,6), \quad (7,7).$$

Here we have $\bar{x} = 4$, $\bar{y} = 5$, and (5) gives

$$m = 14/20 = .7 \quad \text{and} \quad b = 5 - (.7)(4) = 2.2.$$

Thus the least squares line is $y = .7x + 2.2$ as in the preceding section.

To prove the theorem we note that if the sum (4) is expanded the outcome is a quadratic polynomial in m and b. It is not difficult to see that the coefficients of b^2 and mb in this quadratic polynomial are n and $2\Sigma x_i$. The ratio of the second of these to the first is $(2/n)\Sigma\, x_i$, or more simply $2\bar{x}$. So following the method of Problems B45 and B46 at the end of Chapter 2, we apply the transformation

$$b = B - \bar{x}m, \qquad m = m \tag{6}$$

to the sum (4) to get a quadratic polynomial in m and B with no term mB, as in expression (6) in the preceding section.

Let us check the details. The application of (6) to the sum (4) gives

$$\begin{aligned} &\Sigma(mx_i + B - m\bar{x} - y_i)^2 \\ &\quad = \Sigma[m(x_i - \bar{x}) + (B - y_i)]^2 \\ &\quad = m^2\Sigma(x_i - \bar{x})^2 + 2m\, \Sigma(x_i - \bar{x})(B - y_i) + \Sigma(B - y_i)^2. \end{aligned} \tag{7}$$

In the middle term here we note that

$$\Sigma(x_i - \bar{x})(B - y_i) = B\Sigma(x_i - \bar{x}) - \Sigma(x_i - \bar{x})y_i = 0 - \Sigma(x_i - \bar{x})y_i$$

by equation (3). Hence the sum (7) reduces to

$$m^2\, \Sigma(x_i - \bar{x})^2 - 2m\, \Sigma y_i(x_i - \bar{x}) + \Sigma(B - y_i)^2. \tag{8}$$

Here we have three terms, the first two of which involve m but not B, and the third involves B but not m. Hence we can deal with the m and B parts separately in minimizing (8). Now the minimum value of any quadratic polynomial $ax^2 + bx + k$ with constant coefficients a, b, k occurs at $x = -b/2a$ if $a > 0$; this was established in Problem B44 at the end of Chapter 2. Applying this to the first two terms in the sum (8), with m in place of x of course, we see that the minimum occurs by choosing m as formulated in (5) in the theorem.

Finally, to minimize the last term in (8) we write

$$\Sigma(B - y_i)^2 = nB^2 - 2B\Sigma y_i + \Sigma y_i^2.$$

This quadratic polynomial in B is minimized by taking

$$B = \frac{1}{n}\Sigma y_i \quad \text{or} \quad B = \bar{y}.$$

Substituting this into $b = B - \bar{x}m$ in (6) we get $b = \bar{y} - m\bar{x}$, as stated in the theorem.

J1. Prove that $n\Sigma x_i^2 \geq (\Sigma x_i)^2$ for any real numbers $x_1, x_2, \ldots, x_n$ where the sums are over $i = 1, 2, \ldots, n$. Also prove that equality holds iff $x_1 = x_2 = \cdots = x_n$. Suggestion: Use $\Sigma(x_i - \bar{x})^2 \geq 0$.

10.3. The Most Likely Number of Occurrences. If a die is tossed four times in a row, or if four dice are tossed once, what is the most likely number of occurrences of a 6? A plausible guess is one, for the following reason. Since the chance of a 6 on one toss of a die is 1/6, with four tosses it might appear that the chance of getting a 6 exactly once is something like 4/6, or 2/3. Actually it is no such thing.

The probability of getting a 6 exactly once in four tosses of a die is 125/324, or approximately 0.39. Even the probability of getting a 6 *at least once* is only 671/1296 or about .52. This shows that we are more likely to get at least one 6 than none at all. However, the most likely number of occurrences of a 6 is zero. The probability of this is 625/1296, or about 0.48.

The verification of the probabilities cited above is straightforward enough. Since the probability of a 6 on one toss of a die is 1/6, the probability of a 6 four times in a row on four tosses is $(1/6)^4$. The probability of a 6 on none of the tosses is $(5/6)^4$. From elementary probability theory we know that the terms of the expression

$$\left(\frac{5}{6}+\frac{1}{6}\right)^4=\left(\frac{5}{6}\right)^4+4\left(\frac{5}{6}\right)^3\left(\frac{1}{6}\right)$$
$$+6\left(\frac{5}{6}\right)^2\left(\frac{1}{6}\right)^2+4\left(\frac{5}{6}\right)\left(\frac{1}{6}\right)^3+\left(\frac{1}{6}\right)^4$$

are the probabilities of getting a 6 not at all, once, twice, three times, and four times, on four successive tosses. The assertions of the preceding paragraph are easily checked from these numbers. In particular, it is easy to verify that the largest of the five probabilities is $(5/6)^4$, corresponding to the case of getting no 6 in the four tosses.

Now it turns out that there is a simple procedure for deciding the most likely number of successes in a series of trials, without calculating the probabilities in detail. In some situations there is not a single most likely number of successes, but two equally likely cases with highest probability of occurrence. The result is the following theorem.

THEOREM 10.3a. *Suppose that the probability of success of an event is p. In n successive occurrences of the event the most likely number of successes is an integer k defined as follows: if $np + p$ is not an integer, then k is the unique integer lying between $np + p -$ 1 and $np + p$, that is,*

$$np+p-1<k<np+p; \tag{1}$$

on the other hand, if $np + p$ is an integer, then there are two equally likely values of k, namely,

$$k=np+p \quad and \quad k=np+p-1. \tag{2}$$

For example, consider the problem raised at the outset of this section, with a single die tossed four times in a row. The event is one toss of a die; success is getting a 6. Thus $p = 1/6$ and, since the die is tossed four times, $n = 4$. We see that $np + p = 5/6$, which is not an integer. Using (1) we find that the unique value of k is 0, and this is the most likely number of successes.

As a second example, suppose a die is tossed 23 times. What is the most likely number of occurrences of a 6? Again we have $p = 1/6$; but now $n = 23$, and so $np + p = 4$. Here we have the situation where $np + p$ is an integer; so by equations (2) of the theorem, the most likely number of occurrences is 3 or 4. These two possibilities are equally likely.

To prove the theorem, we define q by $q = 1 - p$, so that q is the probability of a failure in one occurrence of the event. The probability of success every time in n repetitions of the event is p^n; the probability of failure every time is q^n. In general, we take it as known from simple probability theory that the terms of the expansion of $(q + p)^n$ give the probabilities of no successes, 1 success, 2 successes, ..., n successes. The terms of the expansion of $(q + p)^n$ are

$$q^n,\ nq^{n-1}p,\ \binom{n}{2} q^{n-2}p^2,\ \ldots,\ \binom{n}{j} q^{n-j}p^j,\ \ldots,\ p^n, \tag{3}$$

where $\binom{n}{j}$ is the binomial coefficient

$$\binom{n}{j} = \frac{n!}{j!(n-j)!}.$$

The term $\binom{n}{j} q^{n-j}p^j$ is the probability of exactly j successes in n repeated occurrences of the event. It is convenient to denote the $n + 1$ terms in (3) by the notation

$$T_0, T_1, T_2, \ldots, T_j, \ldots, T_n. \tag{4}$$

The ratios of successive terms are

$$T_1/T_0,\ T_2/T_1,\ T_3/T_2,\ \ldots,\ T_j/T_{j-1},\ T_{j+1}/T_j,\ \ldots,\ T_n/T_{n-1}, \tag{5}$$

which are easily calculated to be

$$\frac{np}{q},\ \frac{(n-1)p}{2q},\ \frac{(n-2)p}{3q},\ \ldots,\ \frac{(n-j+1)p}{jq},\ \frac{(n-j)p}{(j+1)q},\ \ldots,\ \frac{p}{nq}. \tag{6}$$

We observe that the numerators in the sequence (6) decrease from one term to the next, but the denominators increase. Hence the ratios (6) form a *decreasing* sequence, and similarly for (5). The question then is simply this: Which of these ratios exceed 1, and which are less than 1?

Specifically, suppose that the first j terms of the sequence (5) or (6) exceed 1 and that all subsequent terms are less than 1. It would follow that

$$T_0 < T_1 < T_2 < \cdots < T_j \quad \text{and} \quad T_j > T_{j+1} > \cdots > T_n.$$

Then the largest term in (4) would be T_j. Of course it might happen that some term in (5) equals 1, say $T_j/T_{j-1} = 1$. In this case there would be two largest terms, T_j and T_{j-1}, in the sequence (4).

Consider first the case where no term in (5) or (6) equals 1, and suppose that

$$T_j/T_{j-1} > 1 \quad \text{and} \quad T_{j+1}/T_j < 1.$$

Writing these inequalities in the formulation (6) we get

$$\frac{(n-j+1)p}{jq} > 1 \quad \text{and} \quad \frac{(n-j)p}{(j+1)q} < 1,$$

which reduce to

$$np + p > j \quad \text{and} \quad np + p - 1 < j$$

by simple algebra with the use of $q = 1 - p$. Hence the integer j for which T_j is the largest term in (4) lies between $np + p - 1$ and $np + p$. This integer is called k in the theorem, satisfying the inequalities (1).

Now consider the other possibility, that some term in the decreasing sequence (5) equals 1, say

$$T_j/T_{j-1} = 1 \quad \text{or} \quad [(n-j+1)p]/jq = 1.$$

This gives $j = np + p$, and the sequence $T_1, T_1, \ldots, T_n$ has two largest terms, T_j and T_{j-1}. In the statement of the theorem the symbol k is used to identify these two most likely numbers of occurrences; thus $k = np + p$ and $k = np + p - 1$ as in equations (2).

Example. Define a "successful" toss of three coins as heads on all three. What is the most likely number of successes in 6 tosses of three coins? In 7 tosses?

Since the probability of success in one toss of three coins is 1/8, the answer is no successes in six tosses. In seven tosses, no successes and one success are equally likely, and each of these is more likely than any larger number of successes.

From Theorem 10.3a it is clear that the most likely number of successes in n trials is close to np, where p is the probability of success in one trial. A theorem of Bernoulli says that the number of successes must be close to np in a certain sense. To be specific, let ϵ be any positive number, as small as we please. *Bernoulli's theorem*, or the law of large numbers (not proved here), states that if k is the number of successes in n trials, the probability that

$$np - n\epsilon < k < np + n\epsilon$$

tends to 1 as n increases indefinitely.

At the start of this section we raised a question about probability where our intuition might lead us astray. Another example of this involves what is popularly called the "law of averages," a loosely defined term. A case in point is the following question. Does a basketball player have a better chance of sinking at least two free shots out of four, as against at least one out of two? And is at least three out of six more favorable again? Many people are inclined to answer yes to these questions, on the grounds that the "law of averages" is in our favor with the larger numbers. (Bernoulli's theorem stated above is a kind of mathematical law of averages, but it has no application to the questions under consideration.)

A special case can be stated in terms of coin tossing. Is it more likely that we get heads at least twice in four tosses of a coin, than heads at least once in two tosses? And is it more likely again that we get at least three heads in six tosses of a coin, at least four in eight, and so on? These questions are not difficult to answer, so we set them up in problem J5 below.

J2. What is the least number of tosses of a pair of dice so that the probability exceeds ½ of getting a sum of 7 or 11 at least once?

J3. A die is tossed until there is a repetition of some number, as, for example, in the sequence of tosses 3, 6, 5, 1, 4, 5. What is the smallest number of tosses so that there is at least a fifty-fifty chance of this occurring?

J4. A die is tossed until there is a repetition of the number on the first toss, as for example with the sequence of tosses 3, 6, 5, 1, 4, 5, 1, 3. What is the least number of tosses so that the probability of such a repetition is at least ½?

J5. Let p be the probability that a basketball player can get a basket on one free shot. (i) For what values of p is the player more likely to sink at least two out of four shots than to sink at least one out of two? (ii) Let $P(n, 2n)$ denote the probability of sinking at least n shots out of $2n$. If $p = 1/2$, prove that $P(1, 2) > P(2, 4) > P(3, 6) > \cdots > P(n, 2n) > P(n + 1, 2n + 2) > \cdots$.

J6. *The Birthday Pairing Problem*. Find the smallest number of people required to give a better than fifty-fifty chance that two or more of them have the same birthday. (The year of birth need not match, just the month and day. Ignore February 29 as a birthday; it makes no actual difference in the answer, but it does make the arithmetic more messy if it is included. Also, assume that the other 365 days are equally likely birthdays.)

10.4. Experimental Solutions of Minimal Problems. A geometric solution of Fermat's problem (of finding the point P such that the sum of the distances from P to the vertices of a triangle is a minimum) is given in §9.2. Now we derive another solution of this problem by mechanics, using only the simplest properties of forces. This solution will then be used to suggest experimental solutions of more complicated problems of the same type, problems that cannot be solved by elementary geometric procedures as in the preceding chapter. In the last part of this section we mention briefly the soap-film experiments, one of the most famous of the demonstration procedures for actual visual observation of solutions of minimal distance and minimal surface area problems.

Returning to the Fermat problem, let the given triangle ABC lie in a horizontal plane, with a pulley attached at each vertex. Suppose

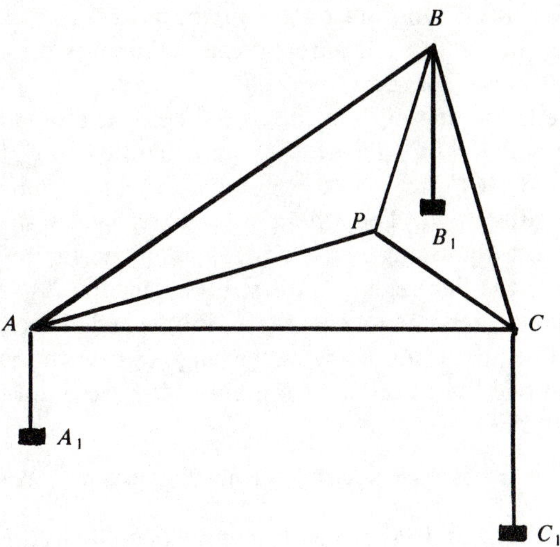

FIG. 10.4a

that three strings are tied at a point P inside the triangle, one string extending over each of the three pulleys at the vertices, as illustrated in Figure 10.4a. Equal weights are attached to the strings at their lower endpoints A_1, B_1, C_1. Thus the line segments PA, PB, PC lie in a horizontal plane, whereas the segments AA_1, BB_1, CC_1 are vertical. The system is released and allowed to reach its own position of equilibrium. We now establish that the point P will coincide with the Fermat point of the triangle, that is, the point such that $PA + PB + PC$ is a minimum.

The weights at A_1, B_1, C_1 will find their natural position in such a way that the center of gravity of the system of weights will come to rest at the lowest possible position. This means that the sum of the distances

$$AA_1 + BB_1 + CC_1 \tag{1}$$

is a maximum. But the total lengths of the three strings is

$$(PA + PB + PC) + (AA_1 + BB_1 + CC_1),$$

and this is a fixed sum. Since the equilibrium position of the system maximizes the sum (1), it automatically minimizes the sum $PA + PB + PC$, as we claimed.

Next we prove that each of the three angles at the point P in Figure 10.4a equals 120°. To see this, we extend each of the line segments AP, BP, CP past P a short distance, as indicated by the dotted lines in Figure 10.4b. Thus there are now six angles at P, equal in pairs. Denote these angles by α, β, γ as in the figure. The three forces along the line segments (or string segments) PA, PB, PC are equal because the same weight, say w, is attached at the end of each string. The equilibrium of the system implies that the forces are in balance. Along the direction of the line PA, for example, the equation of forces is

$$w = w \cos \alpha + w \cos \gamma, \text{ giving } 1 = \cos \alpha + \cos \gamma. \tag{2}$$

For the lines PB and PC the corresponding equations are

$$1 = \cos \alpha + \cos \beta, \qquad 1 = \cos \beta + \cos \gamma. \tag{3}$$

By simple algebra it follows that $\cos \alpha = \cos \beta = \cos \gamma$, and since $\alpha + \beta + \gamma = 180°$ we get $\alpha = \beta = \gamma = 60°$. Hence the angles APB, BPC, and CPA are 120°.

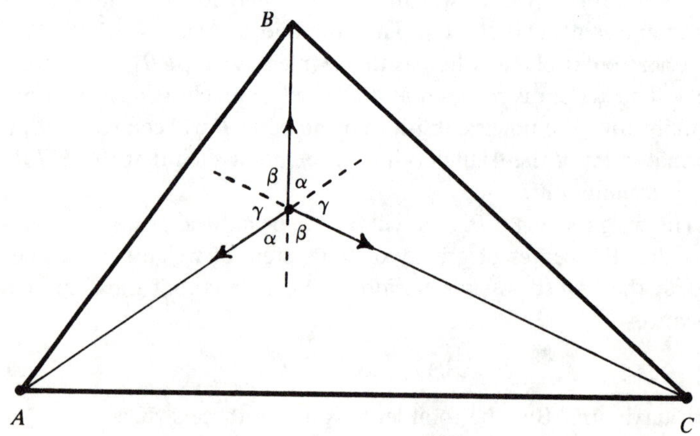

Fig. 10.4b

We have solved the Fermat problem by an application of mechanics. What is of special interest here is the use of the technique above in more general problems. Recall that the Fermat problem might also be called the "airport problem," where we want to locate the airport at a point P so that the sum of the distances to the three towns at A, B, C is a minimum. Now suppose that the towns A, B, C have populations p_1, p_2, p_3, respectively. Consider the problem of the location of the point P so that the total distance traveled by people going to the airport is a minimum. It is logical to assume that the three roads to the airport AP, BP, CP are used in proportion to the size of the populations. Hence the problem is not to locate P so as to minimize the sum of the distances, but to minimize

$$p_1(PA) + p_2(PB) + p_3(PC), \tag{4}$$

because this sum of products takes into account the different amount of traffic on the three roads.

We prove that *the solution of this problem can be obtained by using not equal weights at* A_1, B_1, C_1 *as before, but weights* w_1, w_2, w_3 *that are respectively proportional to the populations* p_1, p_2, p_3. It will make the argument a little simpler if the system of strings, pulley, and weights is constructed so that the three strings emanating from P are equal in length, say length d, so that

$$PA + AA_1 = PB + BB_1 = PC + CC_1 = d.$$

(The argument is valid with strings of different lengths, but a little more complicated.)

The equilibrium position occurs again so that the center of gravity of the weights is at the lowest possible position. Contrasted with the simple formulation (1), this means that the weighted sum

$$w_1(AA_1) + w_2(BB_1) + w_3(CC_1) \tag{5}$$

is a maximum. (This represents the loss in potential energy of the system of weights as they drop from the level of the plane of $\triangle ABC$ to their positions of equilibrium.) But the total sum

$$\begin{aligned} w_1(PA + AA_1) + w_2(PB + BB_1) + w_3(PC + CC_1) \\ = d(w_1 + w_2 + w_3) \end{aligned} \tag{6}$$

is a constant. Hence in the equilibrium position the point P has the property that

$$w_1\,(PA) + w_2\,(PB) + w_3\,(PC) \tag{7}$$

is a minimum, by subtracting (5) from (6). Now if w_1, w_2, w_3 are proportional to p_1, p_2, p_3, then (7) is proportional to (4), and hence the minimum of (4) occurs if P is at the position of equilibrium.

At this point in the argument, the analogy with the Fermat problem in its simple form leads to a difficulty. Where previously we used equations (2) and (3) to prove that $\alpha = \beta = \gamma = 60°$, the calculation of the angles in the present case is more difficult.

The first of equations (2) now takes the form

$$w_1 = w_2 \cos\alpha + w_3 \cos\gamma, \tag{8}$$

and the corresponding equations for the other directions are

$$w_2 = w_1 \cos\alpha + w_3 \cos\beta, \qquad w_3 = w_1 \cos\gamma + w_2 \cos\beta. \tag{9}$$

Also we still have $\alpha + \beta + \gamma = 180°$. But the problem of solving these equations for α, β, γ in terms of w_1, w_2, w_3 does not yield to elementary methods. The problem appears a little simpler perhaps if we write the equations of the equilibrium of forces along directions perpendicular to PA, PB, PC:

$$w_2 \sin\alpha = w_3 \sin\gamma, \quad w_1 \sin\alpha = w_3 \sin\beta, \quad w_1 \sin\gamma = w_2 \sin\beta.$$

These three equations are not independent, because any two imply the third by simple algebra. Nevertheless the basic point remains: elementary methods are not adequate to solve these equations for α, β, γ in terms of w_1, w_2, w_3.

What we can do is to use the physics of the situation to get an experimental solution of the problem. By actually constructing the system consisting of the triangle, the pulleys, the weights, and the strings, we can find the best possible location of P by observing where it comes to rest.

The same observation applies to the extension of the Fermat problem, or the weighted Fermat problem, to a system of more than three points A, B, C. The extension to n points does not yield an elementary mathematical solution, but the experimental solution by mechanics still works.

Soap-film experiments. One of the most fundamental extremal problems is that of finding the surface of least area bounded by a given closed curve in space. The mathematics involved in the general form of this question is difficult in the extreme. Since a soap-film assumes a position to achieve the least surface tension, which in a first approximation corresponds to the least surface area, one approach to the problem is by the experimental procedure of suspending such a film on a wire that is bent into the shape of the desired closed curve.

These questions are beyond the scope of this book, but fortunately there are comprehensive accounts readily available, of which we cite the following: C. Vernon Boys (1956); R. Courant (1940); R. Courant and H. Robbins (1947; pp. 385–397); J. C. C. Nitsche (1974).

10.5. Ptolemy's Theorem. This section contains background material for the next topic, the refraction of light. We will need the following result.

THEOREM 10.5a. *If A, B, C, D are any four distinct points in a plane, then*

$$AB\cdot CD + AD\cdot BC \geq AC\cdot BD, \tag{1}$$

with equality iff (i) *A, B, C, D lie on a circle in that order, or* (ii) *A, B, C, D lie on a straight line with exactly one of B and D lying between A and C.*

The statement that A, B, C, D lie on a circle in that order means that starting from the point A, the succession of points B, C, D lie around the circle in one direction or the other. If we regard a straight line as a circle of infinite radius, condition (ii) is contained in condition (i).

This theorem is an extended version of *Ptolemy's theorem,* which states that *a quadrilateral inscribed in a circle has the property that the product of its diagonals equals the sum of the products of the opposite sides.* We prove here only this basic part of Theorem 10.5a. For the proof in general, we refer the reader to standard works on geometry, listed in the notes at the end of this chapter.

To prove Ptolemy's theorem, let A, B, C, D be four points in that order on a circle. Let 2α, 2β, and 2γ denote the angles subtended at

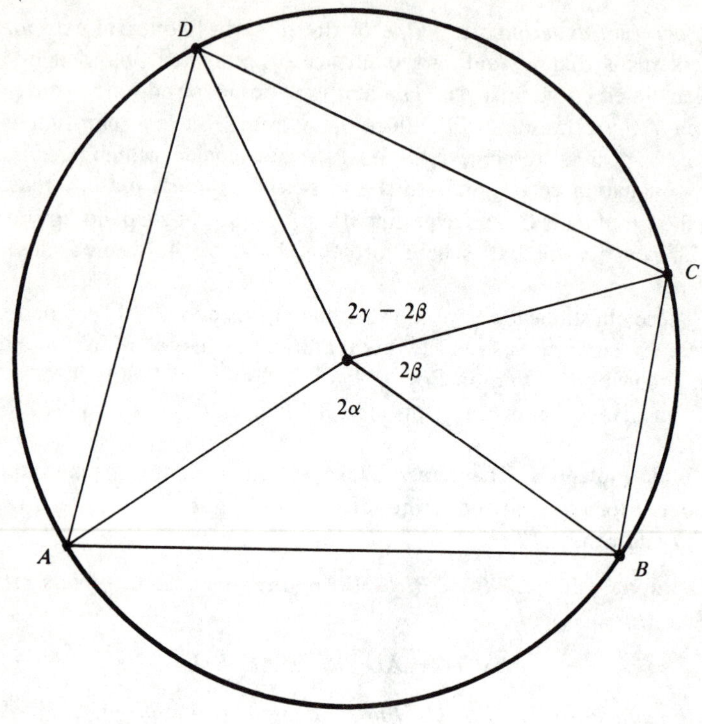

FIG. 10.5a

the center of the circle by the arcs *AB, BC,* and *BD,* respectively, as in Figure 10.5a. Without loss of generality we may take the radius of the circle as the unit of length. Then the length of the chord *AB,* for example, is $2 \sin \alpha$. (Note that this is correct even if 2α exceeds 180°.) The arcs *CD, AD,* and *AC* subtend angles

$$2\gamma - 2\beta, \quad 360° - 2\alpha - 2\gamma, \quad \text{and} \quad 2\alpha + 2\beta,$$

respectively, at the center. The equation we want to establish is $AB \cdot CD + AD \cdot BC = AC \cdot BD$. Replacing the lengths of these line segments by the corresponding sines and dividing by 4, we see that this equation amounts to

$$\sin \alpha \sin (\gamma - \beta) + \sin (\alpha + \gamma) \sin \beta = \sin (\alpha + \beta) \sin \gamma.$$

This is an identity, easily verified by replacing $\sin(\gamma - \beta)$, $\sin(\alpha + \gamma)$, and $\sin(\alpha + \beta)$ by their expanded versions, as in the formulas in §1.2.

Claudius Ptolemy was an astronomer, mathematician, and geographer of the second century A.D. He is best known for his theory that the sun, the planets, and the stars revolve around the earth, a view that was generally accepted for many centuries until Copernicus established about 1530 that the earth rotates on its axis, and the planets travel in orbits around the sun.

J7. If P is any point in the plane of an equilateral triangle ABC, prove that $PA + PB \geq PC$. Under what circumstances is there equality here?

J8. Are there any triangles ABC so that $PA + PB > PC$ holds for every point P in the plane of the triangle? If there are, characterize all such triangles. If there are not, prove it.

10.6. The Refraction of Light. When light travels from one medium into another, for example from air into water, it is refracted in the sense that its path is bent or deflected from travel in a straight line. As illustrated in Figure 10.6a the light travels from A to B via the point P along two straight line segments AP and PB. It is this phenomenon that makes a straight object like a stick appear bent at the water line when partially submerged.

Let the speed of light be v_1 in the medium above the level QR, and v_2 below. With points A and B fixed, the problem is to locate the point P on the line QR to satisfy what is called *Fermat's principle*, or the *principle of least time*. This principle states that the point P is to be located on QR so as to minimize the travel time of the light from A to B via P. Now straight line travel over a distance s at a constant velocity v satisfies the equation $s = vt$, where t is the travel time. Hence we want to locate P on the line QR so that

$$(AP)/v_1 + (PB)/v_2 \tag{1}$$

is a minimum.

Let α be the angle between AP and the perpendicular to QR at P, and β the angle between BP and the perpendicular to QR at P, as shown in Figure 10.6a. We will prove that the solution of the problem is to locate P so that

$$(\sin \alpha)/(\sin \beta) = v_1/v_2. \tag{2}$$

This is known as *Snell's law*. We prove that, if P is located so that this equation holds, then the travel time (1) is a minimum. Thus, if K is any point on QR other than P, we establish that

$$(AP)/v_1 + (PB)/v_2 < (AK)/v_1 + (KB)/v_2. \tag{3}$$

To prove (3), we draw the circle through A, P, and B. Let the perpendicular to QR at P intersect the circle again at the point T, as shown in Figure 10.6b. Applying Theorem 10.5a to the four points A, P, B, T we get

$$AP \cdot BT + AT \cdot PB = AB \cdot PT,$$

with equality because the points lie on a circle. Also we apply Theorem 10.5a to the four points A, K, B, T to get

$$AK \cdot BT + AT \cdot KB > AB \cdot KT.$$

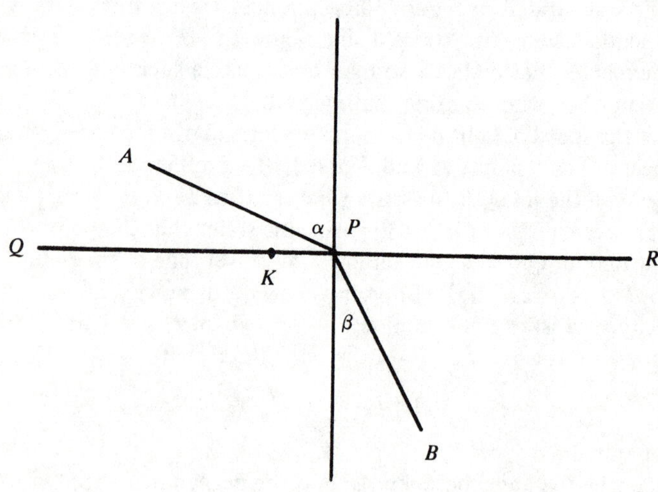

FIG. 10.6a

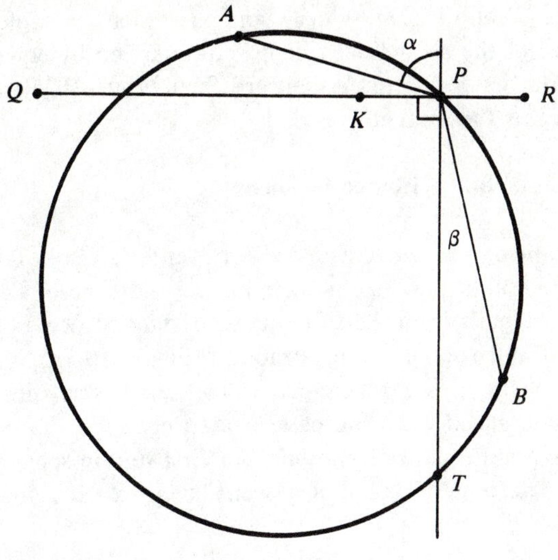

FIG. 10.6b

Here there is a strict inequality because the points $A, K, B, T,$ do not satisfy the conditions for equality in Theorem 10.5a. These two results imply that

$$AP \cdot BT + AT \cdot PB < AK \cdot BT + AT \cdot KB, \tag{4}$$

because $PT < KT$.

Since BT subtends an angle β at P, it follows that BT subtends an angle 2β at the center of the circle, so that $BT = 2r \sin \beta$, where r is the radius of the circle. Similarly, AT subtends an angle $\pi - \alpha$ at P, and hence

$$AT = 2r \sin(\pi - \alpha) = 2r \sin \alpha.$$

It follows that $AT/BT = (\sin \alpha)/(\sin \beta)$. But by assumption the point P is located on the line QR so that $(\sin \alpha)/(\sin \beta) = v_1/v_2$. Thus $AT/BT = v_1/v_2$, and for some constant k we have $AT = kv_1$, $BT = kv_2$. Substituting these into (4), and dividing by k, we get

$$v_2(AP) + v_1(PB) < v_2(AK) + v_1(KB).$$

This gives (3) on division by $v_1 v_2$, and the proof is complete.

The law of the refraction of light, summarized by equation (2), is named after its discoverer Willebrord Snell (1591–1626). His fame is based on this fundamental work.

10.7. Time and Distance Problems.

Least time of descent. Given a point A and a straight line L, determine the point P on L so as to minimize the time of descent on a straight-line path from A to P, where the motion occurs by force of gravity without friction, as illustrated in Figure 10.7a.

After solving this problem we will observe that the solution is readily generalized from the case of a straight line L to a curve on which the point P is to be chosen. Our first step in solving the problem is to determine the time of descent from A to any point P on the straight line L.

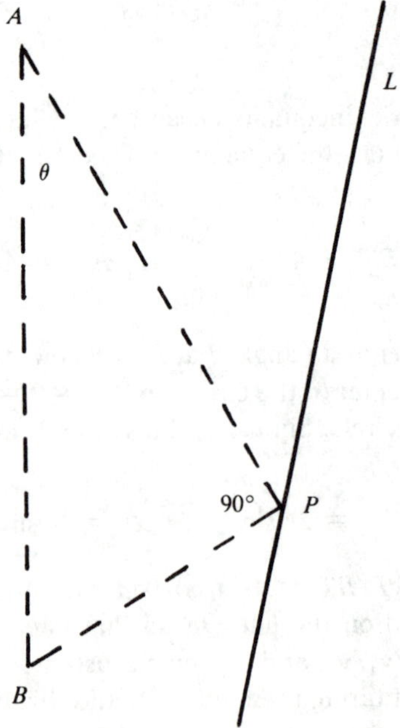

FIG. 10.7a

(The minimum time of descent from A to P, allowing for all possible paths, does not occur along a straight-line path. But this question, known as the "brachistochrone problem" after the name of the curve which provides the solution, is far beyond the scope of this book. It is a problem of considerable historical significance in the calculus of variations.)

Denoting the acceleration of gravity by g, the acceleration along the line AP in Figure 10.7a is $g \cos \theta$, which is the component of g in the direction making an angle θ with the vertical. We assume the standard formula $s = \frac{1}{2}at^2$ from elementary physics, giving the distance s traveled in a straight line in time t of an object starting from rest at time zero and moving with constant acceleration a. In the present case we have $s = AP$ and $a = g \cos \theta$, giving

$$AP = \frac{1}{2}(g \cos \theta)t^2, \qquad t^2 = 2 \sec \theta \cdot (AP)/g. \tag{1}$$

The minimum time t can be obtained by minimizing t^2 in the last equation. This amounts to minimizing $AP \cdot \sec \theta$, since g is a constant. To do this, we locate the point B on the vertical line through A so that angle APB is 90°, as in Figure 10.7a. Note that $AP \cdot \sec \theta = AB$. The location of the point B depends on the location of the point P, as the construction shows. Thus the problem reduces to minimizing the distance AB, as P varies on the line L. The key is to observe that a circle having diameter AB passes through the point P. The smallest circle can now be readily described:

For minimal time of descent from A to P on a line L, the point P should be chosen at the point of tangency of the circle passing through A, tangent to the line L, and having center on the vertical line through A.

This optimal location of the point P is shown in Figure 10.7b, along with the location of the related point B.

This solution depends of the geometric result that the angle APB in a semicircle is a right angle. Since the relationship $AP \cdot \sec \theta = AB$ remains valid if P is moved off the line L to any position on the semicircle shown in Figure 10.7b, it follows from (1) that the time of descent along a straight line path from A to *any* point on the circle is the same as to any other point. For example, the time of descent

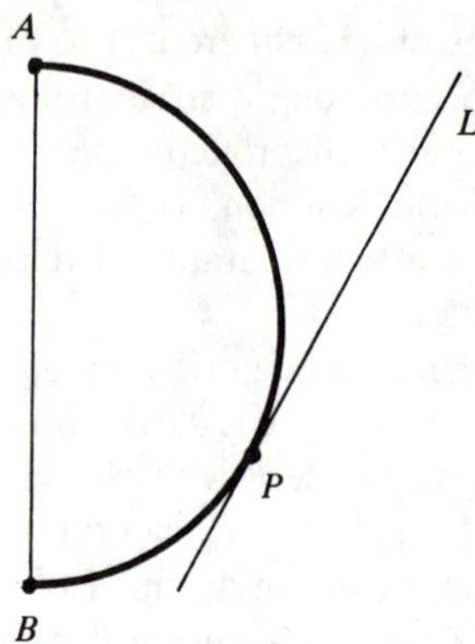

FIG. 10.7b

from A to P is equal to that from A to B. Thus we have proved "Galileo's Theorem" that *the locus of particles sliding along rays emanating from a common point A, with common starting time at A, is at any later time a circle with center on the vertical line through A.*

Moreover, it is clear that the solution above is applicable to any appropriate curve C in place of the line L. The problem is to find the location of the point P on the curve C so as to minimize the time of descent along the straight line AP. The solution is the point P on the curve which is the point of tangency of the smallest circle that reaches the curve C, among all circles passing through A whose centers lie on the vertical line through A. This is illustrated in Figure 10.7c.

J9. Consider the case where the line L is vertical, not passing through A, of course. What should the angle θ be for least time of descent in this case?

Range of a projectile. Consider the question of the maximum horizontal distance traveled by an object, such as a golf ball, starting with an initial speed v_0. If the resistance of the air is neglected, our result is a fairly good approximation for relatively small values of v_0. The approximation is very close to actuality in the case of the athletic event known as the shot-put. Let a and b denote the horizontal and vertical components of the velocity, so that $v_0^2 = a^2 + b^2$. Thus

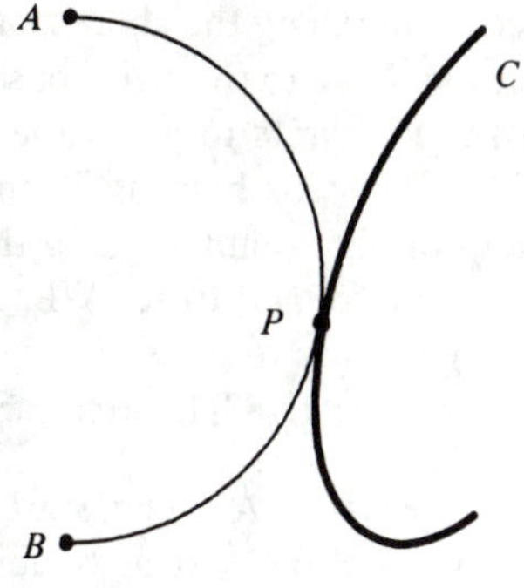

FIG. 10.7c

if the projectile starts its path at an angle θ to the horizontal, then $a = v_0 \cos \theta$ and $b = v_0 \sin \theta$.

The equations of the path are

$$x = at, \qquad y = bt - \frac{1}{2}gt^2,$$

where (x, y) are the coordinates of the point reached at time t. (These equations are rather obvious if we presume the formula $s = \frac{1}{2}at^2$ mentioned earlier in this section. Neglecting gravity the y-coordinate would be $y = bt$, and so the t^2 term is the correction for gravity. As before, g is the acceleration caused by gravity.) Now $y = 0$ for two values of t, namely, $t = 0$ and $t = 2b/g$. Thus the horizontal distance traveled before reaching ground level is obtained from $x = at$ with the second value of t, giving $2ab/g$ as the horizontal range of the projectile.

The problem, then, is to determine the conditions on a and b so as to maximize $2ab/g$. Using the inequality $2ab \leq a^2 + b^2$ we see that the maximum is $(a^2 + b^2)/g$, occurring in case $a = b$. In view of the relation $v_0{}^2 = a^2 + b^2$, we conclude that the maximum horizontal range is $v_0{}^2/g$. This is achieved by starting the flight of the projectile at an angle of 45° to the horizontal, because $a = b$ implies $\cos \theta = \sin \theta$ from the earlier equations, giving $\tan \theta = 1$.

J10. *The lighthouse problem.* A lighthouse stands at a distance 2.5 km. from a shore that is virtually a straight line. A store is located at K, a distance 5 km. along the shore from the point B, which is the closest point on the shore to the lighthouse, as shown in Figure J10. The lighthouse keeper wants to go to the store in the shortest possible time, first by rowing his boat at 3 km. per hour from the lighthouse to some appropriate point P along the shore and then by walking at 5 km. per hour from P to K. What is the least time required to go from the lighthouse to the store, by an optimal choice of the location of P? Suggestion: Use Theorem 5.5a.

J11. *An interception problem.* A cyclist starts from the origin and travels at a constant speed along a straight line, say the positive x axis. At the same time a runner starts from a point P with coordinates (c, y), intending to intercept the cyclist by traveling along a straight line segment from P to the x axis. The runner's speed is half that of the cyclist. Assuming that c is a positive constant, determine the maximum possible value of y as a function of c so that the interception can occur.

J12. A road is to be built from town A to town B. Between the two towns there are two canals, each of which lies between two parallel straight-line sides, although the canals themselves are not parallel to each other, as illustrated in Figure J12. A bridge is to be built over each canal as part of the road, perpendicular to the canal for minimum cost. Determine where the bridges should be built across the canals to obtain the shortest possible road from A to B.

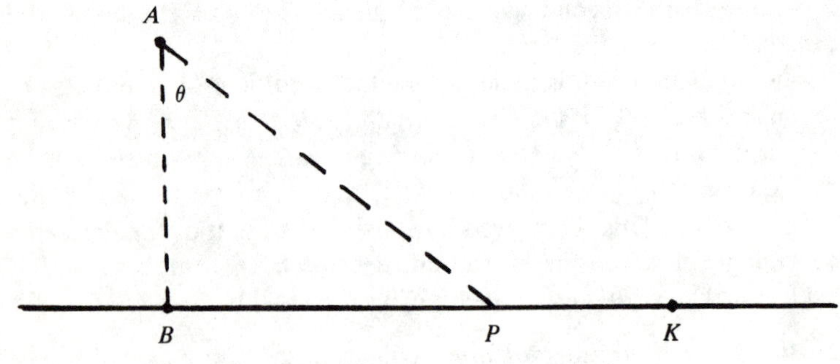

FIG. J10

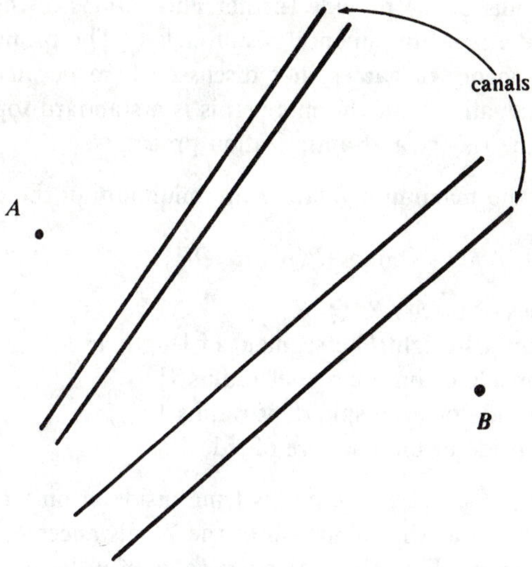

FIG. J12

10.8. Minimax Problems. A *minimax problem* involves finding the maximum value among a set of minima, or vice versa, the minimum value in a set of maxima. For example, consider the following question. Let (a, b) denote the greatest common divisor of the positive integers a and b. Find the maximum value of

$$\min\{(a, b), (a, c), (b, c)\} \tag{1}$$

over all triples of distinct integers a, b, c selected from the set $\{1, 2, 3, \ldots, 100\}$. (For example, if $a = 21$, $b = 42$, and $c = 72$, then $(a, b) = 21$, $(a, c) = 3$, and $(b, c) = 6$, and the value of (1) is 3.

Solution. The answer is 33, from $a = 33$, $b = 66$, and $c = 99$. Suppose, on the contrary, that the minimum value in (1) is some integer $k > 33$. We may assume that the minimum (1) is (a, b), by permuting a, b, and c if necessary. Thus we can write $(a, b) = k$, $(a, c) \geq k$, and $(b, c) \geq k$. Since $3k > 100$, we may take $a = k$ and $b = 2k$. Then $(a, c) \geq k$ becomes $(k, c) \geq k$, and this implies that $(k, c) = k$. Since $c \leq 100$ and $k > 33$, it follows that $c = k$ or $c = 2k$. This is impossible since a, b, and c are distinct.

The problems below provide further illustrations of the minimax concept. There is a fundamental result called "The Minimax Theorem" in the theory of games, not discussed here because it would lead us too far afield. Furthermore, this is a standard topic in textbooks on game theory and optimization processes.

J13. Find the maximum value of the minimum of the distances

$$\min \{PQ, PR, QR\},$$

over all triples of points P, Q, R

(a) lying on a straight line segment of length 1;
(b) lying inside or on a circle of radius 1;
(c) lying inside or on a sphere of radius 1;
(d) lying inside or on a square of side 1.

J14. Let $P_1, P_2, \ldots P_7$ be points lying inside or on a circle of radius 1. Define m as the minimum of the 21 distances P_iP_j between pairs of the points. Find the maximum value of m for all possible locations of $P_1, P_2, \ldots, P_7$.

J15. Let $[a, b]$ denote the least common multiple of positive integers a and b. Determine the maximum value of

$$\min \{[a, b], [a, c], [b, c]\}$$

over all triples of distinct integers a, b, c selected from the set $\{1, 2, 3, \ldots, 100\}$.

10.9. The Jeep Crossing the Desert. A jeep is to travel across a desert but cannot make the trip on one load of fuel. However, if the jeep carries fuel out to storage points in the desert, it is possible to make the crossing. For example, if the jeep can travel 300 miles on one load of fuel, it can cross a desert 400 miles wide on two loads of fuel by depositing 1/3 load of fuel 100 miles out in the desert on a first trip, and then picking up the deposited fuel on a second trip from the starting point.

The problem is to find the minimum amount of fuel needed to cross a desert of a specified width. The solution of this problem is easily attained by starting with the dual question, *How far can the jeep travel using a specified total amount of fuel?* To simplify notation we take the unit of fuel to be the maximum amount that the jeep can carry, and the unit of distance as the distance the jeep can travel on one unit of fuel. Denote by $d(f)$ the maximum distance into the desert from the starting point that the jeep can reach using f units of fuel, by the creation of storage depots in the desert. It is clear that $d(1) = 1$. However, if $f > 1$, we see that $d(f) < f$ because the jeep must return to the starting point at least once for more fuel.

If f is an integer, the solution of the problem is

$$d(f) = 1 + \frac{1}{3} + \frac{1}{5} + \frac{1}{7} + \cdots + \frac{1}{2f - 1}. \tag{1}$$

For $f = 2$ this gives $d(f) = 4/3$, and it may be noted that this result assures us that in the example in the preceding paragraph the maximum possible distance is 400 miles by taking $f = 2$ and the unit of distance as 300 miles.

For a noninteger F, the solution is

$$d(F) = 1 + \frac{1}{3} + \frac{1}{5} + \cdots + \frac{1}{2f - 1} + \frac{F - f}{2f + 1}, \tag{2}$$

where f is the integer part of F; for example if $F = 3.7$, then $f = 3$.

The only assumption used to prove these results is that the jeep reverses direction at only a finite number of points. The results are correct without this assumption, but the proofs are then more complicated. David Gale (1970) makes use of a theorem on integration by Banach in treating this more general case; such methods are beyond the scope of this book. In the case of only a finite number of points where the direction is reversed, Banach's theorem can be stated and proved without calculus, as in the lemma we introduce later.

First we prove that it is possible to travel the distance $d(f)$ as in formula (1), using f units of fuel. Then we establish the more difficult result that this is the maximum distance that can be achieved. The extension to formula (2), where F is not an integer, is left to the reader in the problems below.

In case $f = 1$ it is clear that the jeep can travel a distance 1, and formula (1) gives $d(1) = 1$ accordingly. This is used as a basis for induction on f. Assuming that the distance $d(f)$ as given in (1) can be achieved on f units of fuel, we want to prove the corresponding result with $f + 1$ units of fuel available. We create a fuel depot at the point P located $1/(2f + 1)$ units of distance from the starting point S. This is illustrated in Figure 10.9a where the jeep starts from S, moving to the right initially. Making f trips from S to P, starting with a unit of fuel each time, we can deposit $(2f - 1)/(2f + 1)$ units of fuel on each trip, the rest of the unit of fuel being used to travel the round trip from S to P and back.

This gives us a total deposit of

$$f(2f - 1)/(2f + 1) \tag{3}$$

units ot fuel at the storage depot at P. Now we make trip number $f + 1$ from S to P, starting with the final unit of fuel, arriving with $2f/(2f + 1)$ units of fuel. Adding this to (3) we find that we have arrived at P with exactly f units of fuel available. Starting now from P, we can travel the distance $d(f)$ as in formula (1), on the basis of the induction hypothesis. Since the distance $SP = 1/(2f + 1)$, we see that the distance

$$1 + \frac{1}{3} + \frac{1}{5} + \cdots + \frac{1}{2f - 1} + \frac{1}{2f + 1}$$

can be achieved on $f + 1$ units of fuel.

S P D

1/(2f + 1)

FIG. 10.9a

To complete the proof of formula (1), it must be shown that the distance given is the largest possible using f units of fuel. By way of preparation, we introduce an almost self-evident lemma.

LEMMA. *Consider a finite number of closed intervals, possibly overlapping, on a straight line segment AB of length r. If each point of AB belongs to at least s of the intervals, then the sum of the lengths of the intervals is at least rs.*

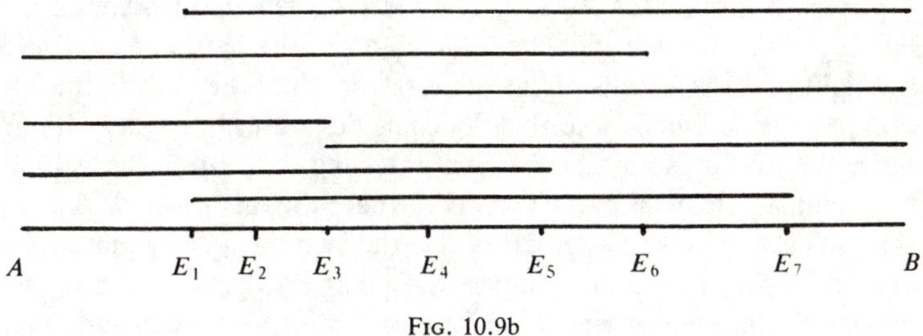

FIG. 10.9b

In Figure 10.9b a simple illustration of this result is given in the case $s = 3$. The intervals, seven in number in the figure, are drawn parallel to and above the line AB, and not on the line segment, for clarity. The endpoints of the intervals are $A, E_1, E_2, \ldots, E_7, B$. Each of the line segments $AE_1, E_1E_2, \ldots, E_6E_7, E_7B$ is covered by at least three of the intervals. Consequently, the total length of the intervals is at least three times that of AB.

With this lemma we are now in a position to prove that the maximum distance that can be traveled on f loads of fuel is given by formula (1), presuming that f is an integer. Consider any path whatsoever that the jeep might take from a starting point S to a destination point D, using f loads of fuel, as shown in Figure 10.9c. The travel path actually lies on the straight line SD, but it is shown schematically in the diagram in a vertically stretched position to make it visible. The short vertical segments are turning points in the path; there is no actual travel along these segments. We prove that

$$SD \leq 1 + \frac{1}{3} + \frac{1}{5} + \cdots + \frac{1}{2f - 1}. \tag{4}$$

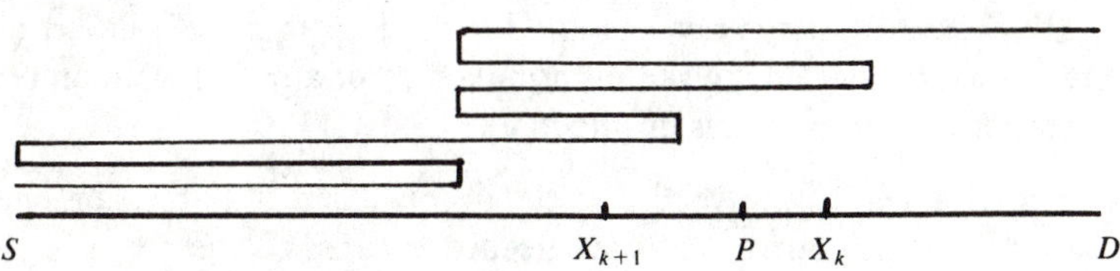

FIG. 10.9c

For every integer $k = 0, 1, 2, \ldots, f$, let X_k denote the point on the line segment SD such that the total length of the path of the jeep to the right of X_k is k units. Thus the total length of the path to the left of X_k is $f - k$ units. Note that X_0 coincides with D, and X_f with S. Since the jeep uses k units of fuel in traveling the path to the right of the point X_k, it must carry k units of fuel past this point. It follows that if P is any point to the left of X_k, the jeep must carry more than k units of fuel past P, and hence the point P is passed by the jeep moving to the right at least $k + 1$ times. From this we conclude that the point P is passed by the jeep moving to the left at least k times, where we are presuming that if P is a point where the jeep reverses its path, the jeep "passes" this point twice as it changes direction. Thus the jeep passes the point P at least $2k + 1$ times in all.

We apply the lemma to the line segment $X_{k+1} X_k$. Any point P in this segment, except perhaps X_k itself, is passed by the jeep at least $2k + 1$ times; so the lemma tells us that the total distance traveled by the jeep between X_{k+1} and X_k is at least $(2k + 1)(X_{k+1} X_k)$. Since by the definition of X_k and X_{k+1} the jeep travels exactly one unit of distance between these points, we conclude that

$$(2k + 1)(X_{k+1} X_k) \leq 1 \quad \text{or} \quad X_{k+1} X_k \leq \frac{1}{2k + 1}. \tag{5}$$

Now the length SD can be separated into its parts,

$$SD = X_f X_{f-1} + X_{f-1} X_{f-2} + \cdots + X_2 X_1 + X_1 X_0. \tag{6}$$

To this we apply the inequality (5) with k replaced successively by $f - 1, f - 2, \ldots, 2, 1, 0$, and the outcome is the inequality (4).

J16. Prove that with F units of fuel, where F is not an integer, the jeep can traverse a desert of the width given in formula (2).

J17. Prove that it is not possible for the jeep to cross a desert of greater width than that given in formula (2), for any path with only a finite number of reversals of direction.

J18. If a jeep can travel 2/3 of the distance across a desert on one load of fuel, how many loads are needed to cross the desert?

Notes on Chapter 10

§10.3. For a more general discussion of the birthday pairing problem, see Mosteller (1965, pp. 46–48).

§10.4. In addition to the references at the end of the section on empirical and experimental mathematics, we cite the work of Almgren and Taylor (1976), Courant and Robbins (1956), and Moulton (1975).

§10.5. For a complete proof of the extended form of Ptolemy's theorem, see Coxeter and Greitzer (1967, pp. 42, 106), or Pedoe (1970, pp. 89–91). It has been pointed out by Apostol (1967) that Ptolemy's theorem can be extended to 3-space. The result is the strict inequality $AB \cdot CD + AD \cdot BC > AC \cdot BD$ for any four distinct points A, B, C, D not lying in the same plane.

§10.6. The neat proof that the refraction of light satisfies Snell's law is adapted from Pedoe (1970, pp. 93–95). Pedoe observes that a number of texts declare that the inequality (3) "cannot be proved by elementary methods." For a very different proof, see Pólya (1954, pp. 149–151).

§10.7. The straight line path of least descent from a point to a circle was posed as Problem A-2 in the 35th Annual William Lowell Putnam Mathematical Competition in December 1974.

The solution of the problem on the maximum range of a projectile follows the arrangement by L. H. Lange (1959) using the inequality of the arithmetic-geometric means, so that the trigonometry is implicit rather than explicit.

§10.8. A basic result in game theory called "the minimax theorem" is given in various textbooks and source books; see for example Ferguson (1967, p. 57).

§10.9. The jeep problem was first solved by N. J. Fine (1947). The solution that we give is adapted from the later paper by David Gale (1970). See Gale's paper for references to other work on related problems, such as the paper on the range of a fleet of aircraft by J. N. Franklin (1960).

Many extremal problems of an applied nature are solved by the use of linear programming and optimization theory, on which there is an extensive literature. We cite the pioneering papers of George B. Dantzig (1951, 1963) and an extensive work by M. R. Hestenes (1975).

A recent book by Pólya (1977) offers a very readable elementary discussion of mathematical techniques in science.

Here is a nice result in applied algebra. If every element a_{ij} of a determinant of order n is a real number satisfying $-1 \leq a_{ij} \leq 1$, then the maximum value of the determinant is at most $n^{n/2}$. This is Hadamard's theorem; the proof is beyond the scope of this book. See Curtis (1977, p. 155).

For an optimization problem of a different type, see the paper by K. R. Rebman, "How to get (at least) a fair share of the cake," in Mathematical Plums, Dolciani Mathematical Expositions, No. 4, Mathematical Association of America, 1979, pp. 22-37.

CHAPTER **11**

EUCLIDEAN THREE-SPACE

11.1. Preliminary Results. Some problems in 3-dimensional space are discussed in Chapter 3, notably Problems C14 to C24 and C26 in §3.4, and Problems C30, C31, C35 at the end of that chapter. These problems do not reveal the complications encountered as we move from the plane to space. Some of the techniques of plane geometry have no counterparts in space and the level of difficulty tends to escalate sharply. For example, although the altitudes of a triangle meet in a point, the analogous result for a tetrahedron fails except in special cases, as the following example shows.

A *tetrahedron* is a natural generalization in 3-space of a triangle in 2-space. It has 4 vertices, not all lying in the same plane, 6 edges joining pairs of vertices, and four triangular faces linking triples of vertices. Consider the tetrahedron with vertices $O(0, 0, 0)$, $A(2, 0, 0)$, $B(2, 1, 0)$, and $C(0, 0, 2)$, as illustrated in Figure 11.1a. The perpendicular from C to the face OAB cuts it at O. The perpendicular from B to the face OAC intersects it at A. The lines CO and BA do not intersect; in fact they are *skew* lines, neither parallel nor intersecting.

One especially useful result that does extend naturally from plane geometry is the distance formula for the distance between the points (x_1, y_1, z_1) and (x_2, y_2, z_2):

$$\sqrt{(x_2 - x_1)^2 + (y_2 - y_1)^2 + (z_2 - z_1)^2}.$$

In this chapter we discuss only a few of the simplest extremal problems in three dimensional geometry. Almost excluded from our consideration are the basic problems about the shortest distance from a point to a line, a point to a plane, and the shortest distance between two skew lines. Although such problems can be solved by the methods developed in earlier chapters, in most cases the best

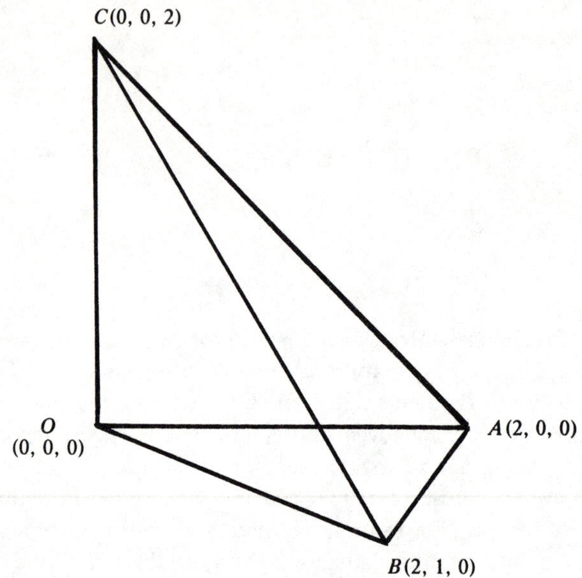

FIG. 11.1a

solution stems from a more detailed study of the geometry of the situation than is developed here. For example, to find the shortest distance from the point (1, 2, 3) to the plane $x - 2y - 2z = 5$ by our methods, we seek the minimum of

$$\sqrt{(x-1)^2 + (y-2)^2 + (z-3)^2}$$

subject to the condition $x - 2y - 2z = 5$. Replacing x by $2y + 2z + 5$ we want to minimize the square root of

$$(2y + 2z + 4)^2 + (y-2)^2 + (z-3)^2.$$

This can be done by the method developed in Problem B45 at the end of Chapter 2, but we do not pursue this because the technique is awkward compared to what can be achieved with coordinate geometry in 3-space.

The shortest distance from a point to a line is an easier matter. For example to find the shortest distance from the point (1, 2, 3) to the line $x = 2 + t, y = 1 + t, z = 2 - t$, we want to minimize the square root of

$$(2 + t - 1)^2 + (1 + t - 2)^2 + (2 - t - 3)^2 = 3t^2 + 2t + 3,$$

by suitable choice of t. By the methods of Chapter 2 we obtain $t = -1/3$ here, giving the shortest distance $\sqrt{8/3}$.

As with plane geometry, some problems are most effectively treated with the use of a coordinate system, and others without.

K1. Find the dimensions of the cone of largest volume that can be inscribed in a sphere of radius 1. Solve the analogous problems in the cases of an inscribed tetrahedron, an inscribed circular cylinder, and an inscribed rectangular solid.

K2. What is the maximum number of points that can be found (i) in a plane, (ii) in 3-space, such that all distances between pairs of points are equal?

11.2. The Isoperimetric Theorem for Tetrahedra. Among all tetrahedra of a given surface area, the regular tetrahedron has the largest volume. An equivalent form of this statement is the following:

THEOREM 11.2a. *Among all tetrahedra of a given volume, the regular tetrahedron has the smallest surface area.*

The proof of this is in two parts, both quite simple. Starting with any tetrahedron T that is not regular, we describe a tetrahedron T_1 having the same volume but smaller surface area. If T_1 happens to be a regular tetrahedron, the proof of the theorem is complete at this stage. On the other hand, if T_1 is not regular we move onto the second stage and prove that the regular tetrahedron with the same volume as T_1 has again smaller surface area. Note that a proof as just outlined involves *no assumption of the existence* of a tetrahedron of least surface area.

Consider any nonregular tetrahedron T. At least one of the triangular faces of T is not an equilateral triangle, because if all four faces are equilateral triangles, the tetrahedron is regular. (In fact, if any three faces of a tetrahedron are equilateral triangles, then it is regular, but we do not need this refinement.) Hence we may assume

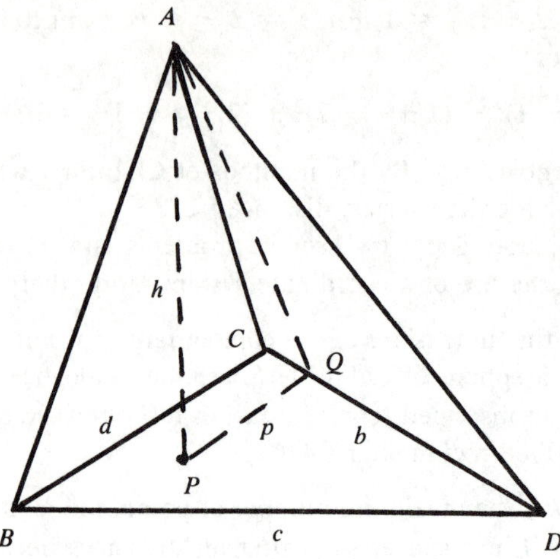

FIG. 11.2a

that the base BCD of T is a triangle with sides of lengths b, c, d, not all equal. Let Δ be the area of this base triangle.

Let AP be the altitude from the vertex A to the base BCD. In Figure 11.2a the point P is shown as inside the triangle BCD, but in general it may be anywhere in the plane of the triangle BCD. If the perpendicular AP has length h, the volume of the tetrahedron T is $h\Delta/3$ by the standard formula.

From the point P we draw three perpendiculars, say of lengths p, q, r to the sides CD, DB, and BC (extended if necessary) of the triangle BCD. In Figure 11.2a only the perpendicular PQ of length p drawn to the side CD is shown, to keep the figure uncluttered. Since $\angle APQ = 90°$, it follows that the length of AQ is $\sqrt{h^2 + p^2}$. From elementary solid geometry we know that AQ is perpendicular to CD. Hence the area of ΔACD is $\frac{1}{2}b\sqrt{h^2 + p^2}$. Using analogous expressions for the areas of triangles ABD and ABC, we see that the surface area S of the tetrahedron T satisfies the equation

$$S - \Delta = \frac{1}{2}\sqrt{b^2h^2 + b^2p^2} + \frac{1}{2}\sqrt{c^2h^2 + c^2q^2} + \frac{1}{2}\sqrt{d^2h^2 + d^2r^2}. \tag{1}$$

To this we apply the inequality of Problem C36 in Chapter 3 to get

$$S - \Delta \geq \frac{1}{2}\sqrt{(bh + ch + dh)^2 + (bp + cq + dr)^2}. \qquad (2)$$

Next, the area of the triangle PCD is $\frac{1}{2}bp$, and similarly the areas of the triangles PBD and PBC are $\frac{1}{2}cq$ and $\frac{1}{2}dr$, respectively. Since the triangle BCD has area Δ, this gives the relation

$$\frac{1}{2}(bp + cq + dr) \geq \Delta \qquad (3)$$

with equality iff the point P lies inside or on the triangle BCD. This with (2) gives the inequality

$$S - \Delta \geq \frac{1}{2}\sqrt{h^2(b + c + d)^2 + 4\Delta^2}. \qquad (4)$$

We construct a tetrahedron T_1 having a base with the same area Δ and having the same volume V as the tetrahedron T, but whose surface area is (say) S_1. We construct T_1 in such a way that $S_1 < S$. The specifications for T_1 with vertices A_1, B_1, C_1, D_1, are as follows: let the base $B_1C_1D_1$ be an *equilateral* triangle of side k chosen so that the area is Δ; let the altitude be h, the same as for the tetrahedron T, so that the volumes of T and T_1 are equal; finally, let T_1 be a right pyramid, that is, so that the altitude from the upper vertex A_1 meets the base $B_1C_1D_1$ at the point P_1 which is the centroid of the base. Thus the perpendiculars from P_1 to the sides of the triangle $B_1C_1D_1$ are of equal length, say p_1, so that

$$\Delta = 3kp_1/2 = \sqrt{3}k^2/4. \qquad (5)$$

The relation analogous to (1) for the surface area S_1 of T_1 is

$$S_1 - \Delta = \frac{3}{2}\sqrt{k^2h^2 + k^2p_1^2} = \frac{1}{2}\sqrt{h^2(3k)^2 + 4\Delta^2}. \qquad (6)$$

But the base of T_1 is an equilateral triangle of area Δ, whereas the base triangle of T has the same area but unequal sides of lengths b, c, d. By the isoperimetric theorem for triangles, the perimeter of the equilateral base is smaller, giving $3k < b + c + d$, and hence $S_1 < S$ by a comparison of (6) and (4). Thus the tetrahedron T_1 has smaller surface area than T, but the same volume. If T_1 happens to be a regular tetrahedron, the desired conclusion follows. If not, we proceed to the second stage of the proof.

Part 2 of the proof. If T_1 is not a regular tetrahedron, we prove that a regular tetrahedron of the same volume as T_1 has smaller surface area. Consider a regular tetrahedron with edges of length r having the same volume as T_1. Calculating the volumes of the two tetrahedra from the formula

$$V = \frac{1}{3}(\text{area of base})\,(\text{altitude}),$$

we obtain

$$\sqrt{2}r^3/12 = \sqrt{3}k^2h/12, \quad 2r^6 = 3k^4h^2. \tag{7}$$

For future reference we note that $k \neq r$. For if $k = r$, then (7) implies that $3h^2 = 2r^2$, and it follows that h is equal to the altitude of a regular tetrahedron of edge r. Thus T_1 would be a regular tetrahedron, contrary to the hypothesis of this part of the proof.

The surface area of a regular tetrahedron with edge r is $\sqrt{3}r^2$; we want to prove that this is less than S_1. Using (6) and (5), this means that we want to prove

$$\sqrt{3}r^2 < \sqrt{3}k^2/4 + \frac{1}{2}\sqrt{9h^2k^2 + 3k^4/4}.$$

Multiplying by $4/\sqrt{3}$ and moving a term, we want to establish

$$4r^2 - k^2 < \sqrt{12h^2k^2 + k^4}.$$

If $4r^2 - k^2 \leq 0$, this result is obvious. Otherwise $4r^2 - k^2 > 0$ and we can square both sides to get the equivalent inequality

$$16r^4 - 8r^2k^2 + k^4 < 12h^2k^2 + k^4. \tag{8}$$

From (7) we have $12h^2k^2 = 8r^6/k^2$, so that (8) reduces to

$$2r^4 < r^2k^2 + r^6/k^2. \tag{9}$$

This result follows at once from $2xy < x^2 + y^2$ with strict inequality because the substitutions $x = rk$ and $y = r^3/k$ give $x = y$ iff $r = k$, whereas we have already observed that $r \neq k$.

K3. (i) Give an example to show that it is not true that the sum of the lengths of *any* four edges of a tetrahedron exceeds the sum of the lengths of the other two. (ii) However, prove that the sum of the lengths of any two opposite edges is less than the sum of the lengths of the other four edges, for example, that $AB + CD < AC + AD + BC + BD$ in a tetrahedron $ABCD$. (iii) Prove analogous results for the squares of the edges.

K4. In any tetrahedron prove that the sum of the areas of any three faces exceeds that of the fourth.

K5. Prove that the sum of the distances from any point P in the interior of a regular tetrahedron to the four faces is independent of the location of P.

11.3. Inscribed and Circumscribed Spheres of a Tetrahedron. Any tetrahedron has a circumscribed sphere that passes through the four vertices, and also an inscribed sphere having each of the faces of the tetrahedron as a tangent plane.

THEOREM 11.3a. *If R and r are the radii of the circumscribed sphere and the inscribed sphere of a tetrahedron, then $R \geq 3r$, with equality iff the tetrahedron is regular.*

This is the generalization to 3-space of Euler's result $R \geq 2r$ for the radii of the circumscribed circle and the inscribed circle of a triangle, given in Theorem 9.4a. The proof is a fairly straightforward generalization, as follows.

Given any tetrahedron $ABCD$ it is possible to impose a system of axes so that the coordinates of A, B, C, and D are

$$(0, 0, 0), \quad (3a, 0, 0), \quad (3b, 3c, 0) \quad \text{and} \quad (3d, 3e, 3f), \qquad (1)$$

where a, b, c, d, e, f are appropriate real numbers. The factor "3" can be used here without loss of generality, for later convenience. Let D_1, C_1, B_1, A_1 be the centroids of the triangles ABC, ABD, ACD, BCD, respectively. The coordinates of these centroids are

$$(a + b, c, 0), \quad (a + d, e, f), \quad (b + d, c + e, f),$$

$$(a + b + d, c + e, f),$$

by simple averaging of the coordinates of the vertices of each triangle. The line segments AB and A_1B_1 are parallel and of lengths $3a$ and a, respectively. Similarly the segments BC, BD, CD are parallel to B_1C_1, B_1D_1, C_1D_1, respectively, and the lengths are in the ratios 3 to 1 in each case. Hence the tetrahedra $ABCD$ and $A_1B_1C_1D_1$ are similar with corresponding lengths in the ratio 3 to 1. The radii of the circumscribed spheres of these tetrahedra are also in the ratio 3 to 1, say radii R and $R/3$. But the inscribed sphere of the tetrahedron $ABCD$, with radius r, is the smallest sphere that intersects all four faces of the tetrahedron. Hence $R/3 \geq r$, or $R \geq 3r$, with equality iff the inscribed sphere of $ABCD$ is identical to the circumscribed sphere of $A_1B_1C_1D_1$. This property is readily seen to hold in the case of a regular tetrahedron $ABCD$.

Conversely, we now assume that the inscribed sphere of $ABCD$ coincides with the circumscribed sphere of $A_1B_1C_1D_1$, and prove that $ABCD$ is a regular tetrahedron. To do this we look at what the assumption means in the plane of the triangle $A_1B_1C_1$, that is, the plane $z = f$. It means that the circumscribed circle of triangle $A_1B_1C_1$ is identical with the inscribed circle of the triangle T formed by the intersection of the plane $z = f$ with the tetrahedron $ABCD$. The triangle T has vertices

$$(d, e, f), \quad (2a + d, e, f), \quad (2b + d, 2c + e, f), \tag{2}$$

as is easily calculated. The midpoints of the sides of this triangle are precisely the points A_1, B_1, C_1. This implies, as in the proof of Theorem 9.4a, that the triangle T is equilateral. Subtracting d, e, f, respectively, from the coordinates of the vertices of T in (2), this implies that

$$(0, 0, 0), \quad (2a, 0, 0), \quad (2b, 2c, 0) \tag{3}$$

are the vertices of an equilateral triangle. Hence the triangle ABC, with coordinates in (1) exactly 3/2 times these coordinates (3), is also equilateral. Similarly, the other triangle faces of $ABCD$ are equilateral, and so the tetrahedron $ABCD$ is regular.

11.4. Shortest Paths on a Sphere. Among the circles that can be drawn on a sphere, a *great circle* is one whose radius is the same as that of the sphere itself. Assuming the earth to be a smooth sphere, which is a good approximation, the lines of longitude running through the north and south poles are great circles, whereas the only line of latitude that is a great circle is the equator. Through any two *antipodal points* on a sphere, i.e., points which are diametrically opposite, there are infinitely many great circles. Any other pair of points P and Q determine one and only one great circle, and *the minor arc PQ of this great circle is the shortest distance from P to Q on the surface of the sphere.* The proof of this assertion is the goal of this section. At the same time we prove that the shortest distance between two antipodal points on the sphere is half the length of a great circle.

First we establish a preliminary result that is of interest in itself.

THEOREM 11.4a. *At any vertex of a tetrahedron, the sum of any two face angles exceeds the third. In detail, if BAC is the largest face angle at the vertex A in a tetrahedron $ABCD$, then*

$$\angle BAD + \angle CAD > \angle BAC. \tag{1}$$

We establish first that without loss of generality we may assume that $\angle ABD = \angle ABC$. In case these angles are not equal, equality can be obtained by moving the point C into a new position along the half-line AC, the motion being toward A or away from A as necessary. (As is readily observed from Figure 11.4a, the angle ABC decreases as C is moved toward A, but increases if C is moved away from A.) Note that such a relocation of C does not disturb the angles involved in the inequality (1).

With $\angle ABD = \angle ABC$ thus obtained, we choose the point E on BC so that $\angle BAD = \angle BAE$, as illustrated in the figure. The triangles ABD and ABE are congruent, implying that $BD = BE$ and $AD = AE$. By the triangle inequality it follows that $BD + DC > BC$, from which we conclude that $DC > EC$. Now compare the triangles ADC and AEC; we have $AD = AE$ and a common side AC. It follows, by an application of the law of cosines, that the angle opposite DC in triangle ADC exceeds the angle opposite EC in triangle AEC. Thus we have $\angle DAC > \angle EAC$, and this proves (1).

We now use this theorem to prove the main result of this section, that the shortest path between two points on a sphere is along an arc of a great circle.

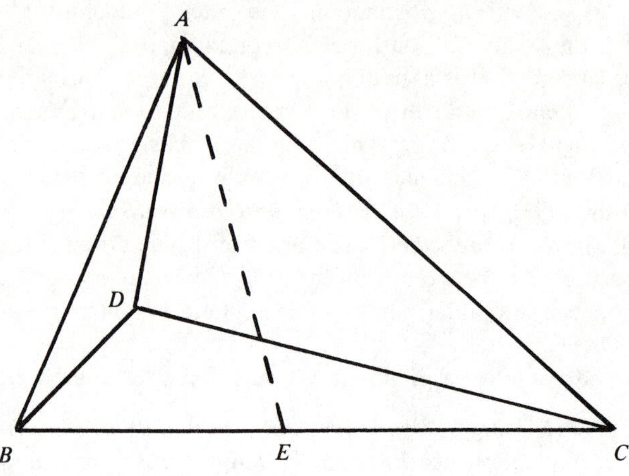

FIG. 11.4a

THEOREM 11.4b. *Let P and Q be any two points on the surface of a sphere. Let Arc PQ (note the capital A in Arc) denote the length of the minor arc of the great circle from P to Q, or in the case of antipodal points on the sphere, let Arc PQ denote half the length of a great circle. If arc PQ (with a small a in arc) denotes the length of any other arc from P to Q, not a great circle, then Arc PQ < arc PQ.*

We may assume that the sphere has radius 1. Let K be a fixed point on the arc PQ, but not on the great circle Arc PQ, as shown in Figure 11.4b. (We use the term "arc PQ" to mean either the length of the arc or the actual arc itself. The usage intended is clear from the context. Similarly for the term "Arc PQ.")

First we show that

$$\text{Arc}\, PK + \text{Arc}\, KQ > \text{Arc}\, PQ. \tag{2}$$

Let O denote the center of the sphere. In the tetrahedron $OPKQ$ we know from Theorem 11.4a that

$$\angle POK + \angle KOQ > \angle POQ.$$

Since the length of an Arc XY on a great circle is proportional to the angle XOY it subtends at the center, the inequality (2) follows.

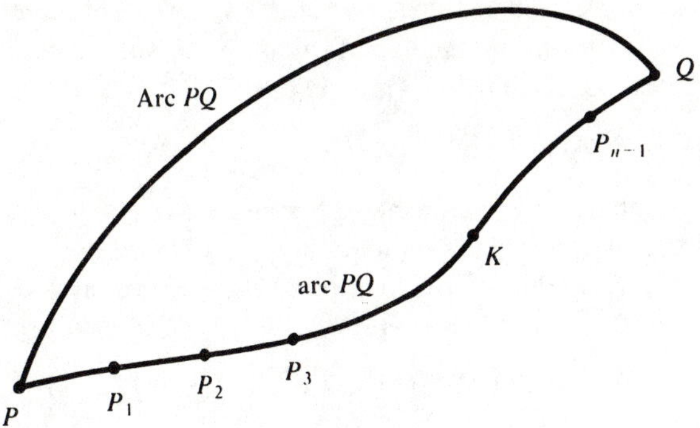

FIG. 11.4b

Next, define the positive number β by the equation

$$\beta(\text{Arc}\,PK + \text{Arc}\,KQ) = \text{Arc}\,PQ, \tag{3}$$

so that $\beta < 1$. Choose a positive angle θ sufficiently small so that $\cos\theta > \beta$. This can be done because $\cos\theta$ tends to 1 as θ tends to 0.

Now if A and B are any two points on the sphere that are sufficiently close so that the great circle Arc AB subtends an angle $\leq \theta$ at the center of the sphere, then we prove that

$$AB > \beta(\text{Arc}\,AB), \tag{4}$$

where AB denotes the straight-line distance from A to B inside the sphere. To prove (4) we use Theorem 6.5a. This gives $\tan\theta > \theta$ for any acute angle θ, which implies that $(\sin\theta)/\theta > \cos\theta$. If Arc AB subtends an angle 2λ at the center of the sphere, where $2\lambda \leq \theta$, we see that $AB = 2\sin\lambda$, Arc $AB = 2\lambda$, and $AB/(\text{Arc}\,AB) = (\sin\lambda)/\lambda > \cos\lambda > \cos\theta > \beta$, and this proves (4).

Now choose a succession of points $P_1, P_2, \ldots P_{n-1}$, one of which is taken to be K, along the arc PQ so that each of the straight line segments

$$PP_1, P_1P_2, P_2P_3, \ldots, P_{n-1}Q \tag{5}$$

subtends an angle $\leq \theta$ at the center of the sphere. This can be done by taking n sufficiently large. Let S_1 denote the sum of the straight line distances (5), so that by (4) we have

$$S_1 > \beta(\text{Arc } PP_1 + \text{Arc } P_1P_2 + \cdots + \text{Arc } P_{n-1}Q).$$

The sum of the terms in parentheses here is $\geq$ Arc PK + Arc KQ by repeated use of the "triangle inequality" (2). (Here use is made of the fact that K is one of the points in the succession $P_1, P_2, \ldots P_{n-1}$.) Using this with (3) we get

$$S_1 > \beta(\text{Arc } PK + \text{Arc } KQ) = \text{Arc } PQ. \tag{6}$$

Finally, we see that the arc PQ is separated by the points $P_1, P_2, \ldots, P_{n-1}$ into n pieces such that

$$\begin{aligned} \text{arc } PQ &= \text{arc } PP_1 + \text{arc } P_1P_2 + \cdots + \text{arc } P_{n-1}Q \\ &> PP_1 + P_1P_2 + \cdots + P_{n-1}Q = S_1. \end{aligned} \tag{7}$$

Results (6) and (7) imply that arc $PQ >$ Arc PQ.

K6. Is there an inequality analogous to Theorem 11.4a for the dihedral angles (defined in the parenthetical remark at the end of the problem) between pairs of the faces meeting at a vertex of a tetrahedron? Specifically, for the three dihedral angles between pairs of faces ABC, ABD, and ACD in a tetrahedron $ABCD$, is it true that the sum of any two of the angles exceeds the third? (The *dihedral angle* between two intersecting planes is the angle PQR, where Q is a point on the line of intersection and the point P lies on one plane and R on the other such that the lines PQ and RQ are perpendicular to the line of intersection of the planes.)

Miscellaneous Problems.

K7. The largest rectangular packages that can be mailed in the U.S.A., according to postal regulations, have length plus girth not more than 84 inches, where "girth" means the distance around the package in the two shorter dimensions. (Thus if the dimensions are x by y by z inches, where $x \leq y \leq z$, then z is the length and $2x + 2y$ is the girth.) Find the dimensions of the package of greatest volume that the U.S. mail service accepts.

K8. Consider a rectangular room 20 feet long, 10 feet wide, and 10 feet high. A spider is located on an end wall (dimensions 10 by 10) at a point equally distant from the side walls, one foot down from the ceiling. The spider decides to investigate a fly on the opposite end wall at a position also midway between the sidewalls, one foot up from the floor. It is clear that the spider can walk to the fly along a 30-foot path by going up to the ceiling, along the center line of the ceiling, and then down the other end wall. (i) Is this the shortest path the spider can walk, allowing for all possible routes along the walls, the ceiling, and the floor? Solve the problem also in the cases where the room is (ii) 15 feet long and (iii) 25 feet long.

K9. Find the shortest distance from the point $(4, 0, 0)$ to the point $(2\sqrt{2}, 2\sqrt{2}, 7)$ along the surface of the cylinder $x^2 + y^2 = 16$. (*Remark*: Since z is missing from the equation, this means that the surface consists of all points (x, y, z) satisfying the condition $x^2 + y^2 = 16$, so that z is arbitrary.)

K10. Find the shortest distance from the point $(2, 0, \sqrt{5})$ to $(-2, 0, \sqrt{5})$ along the surface of the cone $4z^2 = 5x^2 + 5y^2$.

K11. *The extension of Heron's theorem to 3-space.* Consider a line CD and two points A, B not on the line, with the points A, B, C, D not coplanar, i.e., not lying in a plane. Where should the point P be located on the line CD in order to minimize $AP + PB$? This question can be formulated in terms of coordinates as follows. Take the x axis to be the line CD, with the line through A perpendicular to CD chosen as the y axis, so that the origin is the intersection point of this perpendicular with CD. Let $(0, a, 0)$ be the coordinates of A, where we may assume $a > 0$ by proper choice of the positive direction on the y axis. Then the point B has coordinates (b, c, d), say, where $c^2 + d^2 > 0$, because the point B does not lie on the x axis. Assign the coordinates $(x, 0, 0)$ to P, and the problem is to determine x in terms of a, b, c, d to minimize $AP + PB$. *Suggestion*: Use the theory in §3.5.

K12. A wooden cube is to be cut into 27 equal subcubes. Find the minimum number of cuts to achieve this, assuming that after the first cut the pieces can be stacked so that subsequent cuts can go through two or more pieces of wood.

K13. Consider a wedge cut from the circular cylinder $x^2 + y^2 = r^2$ by the parallel planes $z = 0$ and $z = h$, and by two half-planes intersecting in the z axis. Thus if θ is the angle between the two half-planes, the volume of the wedge is $\theta r^2 h/2$ and its surface area is $2rh + \theta rh + \theta r^2$. Taking r as a constant, find conditions on h and θ to maximize the volume of the wedge for a fixed surface area.

K14. Find the shortest distance from the origin to the surface $z = 8/x^2 y$.

K15. Let n points, not necessarily distinct, be chosen on the surface of a sphere of radius 1 so as to maximize the sum of the squares of the $n(n-1)/2$ distances between pairs of points. Prove that the maximum sum is n^2. (This is a variation on Problem A4 of the 29th William Lowell Putnam Mathematical Competition, used here with the permission of the Mathematical Association of America. It is rather surprising that the maximum here is no larger than for a unit circle, as in problem F4 of §6.5. The additional "elbow room" provided by a sphere avails us nothing extra. However, for the sum of the $n(n-1)/2$ distances between pairs of points, the maximum is larger for a sphere than for a circle, for example, in the case $n = 4$.)

K16. Find the maximum value of the sum of the straight-line distances $AB + BC + CA$, where A, B, and C are points on the surface of a sphere of radius r.

K17. In the preceding problem, find the maximum value of the sum Arc AB + Arc BC + Arc CA, where Arc AB denotes the shortest distance from A to B along the surface of the sphere.

Notes on Chapter 11

§11.2. No assumption is made here of the existence of a solution of the isoperimetric problem for tetrahedra. Such an assumption is made tacitly in some proofs in the literature.

This isoperimetric theorem is just a special case of a more general result for convex polyhedra by L. Lindelöf; details can be found in L. Fejes Tóth (1964, p. 283).

The inequality $AB \cdot CD + AC \cdot BD > AD \cdot BC$ holds for any tetrahedron $ABCD$; see Apostol (1967).

More extensive discussions than in problems K8, K9, K10 of shortest distances along surfaces can be found in Pólya (1954, pp. 161–167) and Tietze (1965, Chapter 2, "Travelling on Surfaces").

Problem K12 was suggested by Scott Niven, K13 by Calvin T. Long.

Here is a generalization to 3-space of a problem solved in §3.4: Given a trihedral angle and a point P in its interior, find a plane through P to form a tetrahedron of minimum volume. The solution is the plane intersecting the trihedral angle in a triangle having P as its centroid; see problem E 1394, Amer. Math. Monthly, 67 (1960), pp. 595–596.

Here is an interesting problem with an unexpected answer: How should eight points be placed on a sphere to maximize the volume of the polyhedron formed by those eight vertices? The inscribed cube comes to mind as a possibility here, with volume $8\sqrt{3}\,r^3/9$ where r is the radius of the sphere. A double pyramid on a regular hexagon (on the sphere $x^2 + y^2 + z^2 = r^2$, two vertices at $(0, 0, \pm r)$ and six vertices forming a regular hexagon on the circle $x^2 + y^2 = r^2$, $z = 0$) gives a larger volume, namely $\sqrt{3}\,r^3$. But the optimal arrangement is not so regular as the cube or the double pyramid. For details of the location of the eight points to give maximum volume, see the solution by Berman and Hanes (1970).

Although calculus is not used overtly in this chapter, it is present in a hidden form because the formula for the volume of a tetrahedron cannot be derived without appealing to a limiting process. For an instructive statement about this result, in the larger setting of Hilbert's third problem, see the review by Ross Honsberger of a book by V. G. Boltianskii in the Bulletin of the Amer. Math. Soc., new series, vol. 1 (1979), pp. 646–650.

CHAPTER **12**

ISOPERIMETRIC RESULTS NOT ASSUMING EXISTENCE

12.1. The Need for a Closer Study. In contrast with the proofs of the isoperimetric theorems in Chapter 4, we now give arguments without the assumption that a solution exists. The new idea introduced here is the "inner parallel polygon," in the next section. This is used to prove the isoperimetric theorem in §12.3, followed by the isoperimetric result for polygons.

In §4.3 we established that if C_1 is any simple closed curve other than a circle, there is another curve C_2 of the same perimeter that encloses a larger area. Can we conclude at once from this that the circle encloses the largest area? To show that we cannot, we give a neat argument by Oscar Perron that has already been used in §4.5. Given any positive integer n, except 1, note that n^2 is a larger integer. Thus given any positive integer other than 1, there is another integer that is larger. But of course we cannot conclude that 1 is the largest integer.

A geometric argument, somewhat more sophisticated than Perron's, has been given by Rademacher and Toeplitz (1957, p. 140), to show the need to establish the existence of a solution in some way. These authors also give, on pages 142–146, a proof of the isoperimetric theorem assuming existence, and at the end of the argument they stress that such a proof is incomplete. They comment that the complete solution would require the "systematic building up of an extensive theory," which they prefer not to do. However, with the introduction of the inner parallel polygon, no extensive theory is needed to formulate an argument that is complete and at the same time not very difficult.

The conjecture that the circle encloses the largest area is plausible from the start, for most people. This being so, why is the isoperimetric problem so famous? Kelly and Weiss (1979, p. 243) answer that the interest lies in the challenge of the subtle and difficult existence question. In the absence of a proof unclouded by any logical gap, can we be certain of our intuition?

12.2. The Inner Parallel Polygon. Starting with any convex n-gon P, we now describe a geometric construction having powerful consequences. Shrink the polygon P by moving each of its sides inward, each side moving parallel to itself, and all sides moving inward at the same rate of speed. As the polygon P shrinks, its vertices move along the bisectors of its angles. The process stops when for the first time one or more of the sides of P shrinks to a point. This can happen with the polygon P contracting in three different ways, into another polygon with fewer sides, into a line segment, or into a single point.

First, consider the case where the convex polygon P shrinks to a polygon P_1, called the *inner parallel polygon* of P. This is illustrated in Figure 12.2a where the polygon P is $BCDEF$, and the inner parallel polygon P_1 is $B_1C_1D_1E_1$, the side BF having shrunk to the

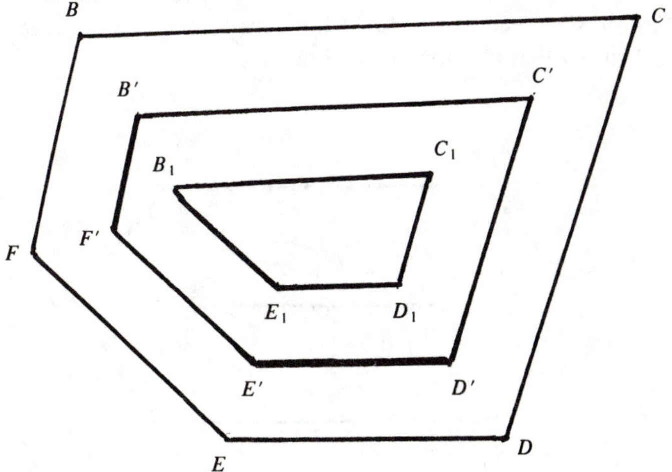

FIG. 12.2a

point B_1. The polygon $B'C'D'E'F'$ in the figure illustrates an intermediate stage in the contraction process.

A second outcome is that the polygon P shrinks to a straight line segment. This is illustrated in Figure 12.2b, where the polygon P is $BCDEFG$, which shrinks to the line segment B_1C_1. The vertices B, G, F, have coalesced into B_1, and the vertices C, D, E into C_1.

A third outcome is that the polygon P shrinks to a point, with all the vertices of P coalescing to this single point. This happens iff P is a polygon circumscribing some circle. The polygon shrinks to the center of this circle.

We now analyze the relationship of the areas and the perimeters of the original polygon P and the inner parallel polygon P_1. Let r denote the distance each side of the polygon P moves before the shrinking process stops at the inner parallel polygon P_1, or at the degenerate forms of a line segment or a point. From the vertices of P_1 we draw perpendiculars, each of length r, to the nearby sides of the original polygon P. This is illustrated in Figure 12.2c, which is just Figure 12.2a with further detail.

The polygon P is thus dissected into other polygons as follows:

(1) the inner parallel polygon P_1; this is $B_1C_1D_1E_1$ in Figure 12.2c;

(2) rectangles based on the sides of P_1; there are four such rectangles in the figure, because P_1 has four sides; each of these rectangles has a pair of opposite sides of length r;

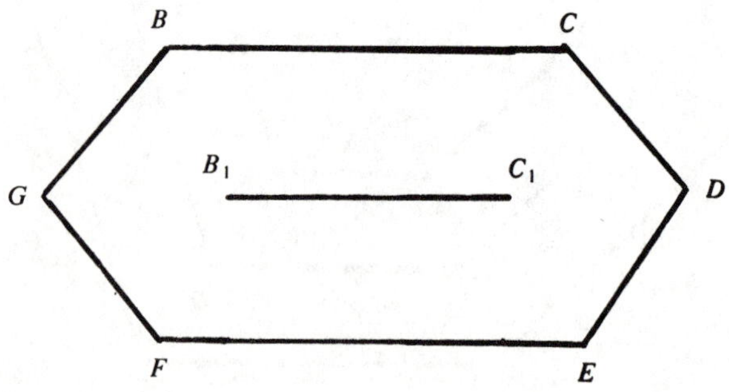

FIG. 12.2b

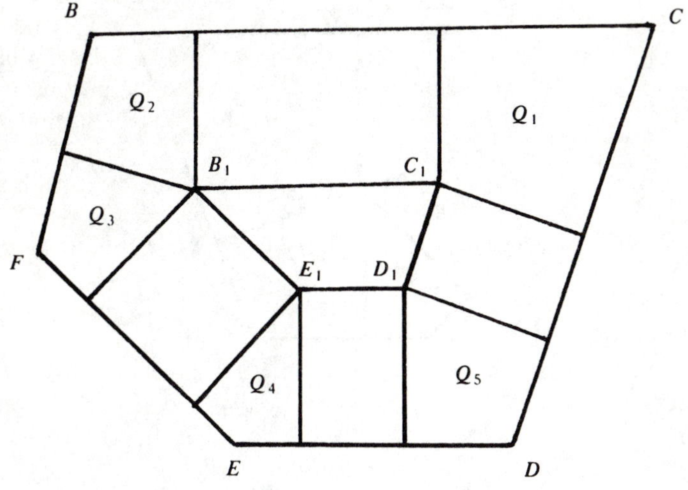

FIG. 12.2c

(3) n kite-shaped quadrilaterals, one at each vertex of P; these are labeled Q_1, Q_2 Q_3, Q_4, Q_5 in Figure 12.2c.

These kite-shaped quadrilaterals can be fitted together, as illustrated in Figure 12.2d, to form an n-gon denoted by P^*. This polygon circumscribes a circle of radius r.

Based on this dissection of the n-gon P, we now observe that

$$A = A_1 + rL_1 + A^*, \quad L = L_1 + L^*, \tag{4}$$

where A, A_1, and A^* denote the areas, and L, L_1, and L^* the perimeters, of the polygons P, P_1, and P^*, respectively. The three terms A_1, rL_1, and A^* in the first equation correspond precisely to the parts (1), (2), and (3) of the dissection above. The equation $L = L_1 + L^*$ follows from the fact that as the polygon P shrinks to P_1, the loss in perimeter is precisely the portions of the sides of P that are used to create the polygon P^*.

Next we derive equations analogous to (4) for the case where the inner parallel polygon degenerates into a straight line segment, say of length d. This is the line segment B_1C_1 in the illustration in Figure 12.2e, which is an elaboration of Figure 12.2b. Each of the

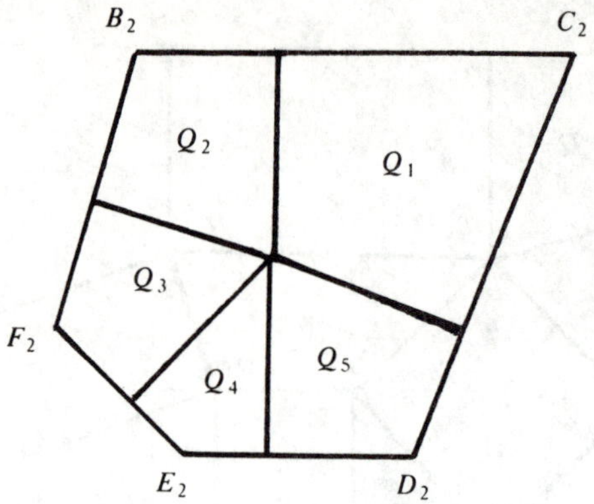

FIG. 12.2d: The Polygon P^*

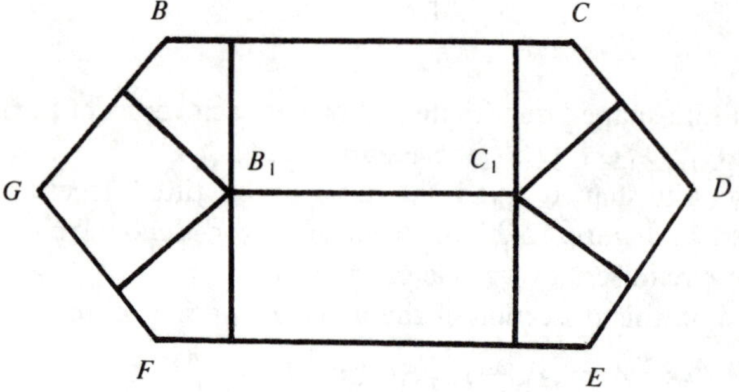

FIG. 12.2e

perpendiculars from B_1 and C_1 to the nearby sides of the polygon P has length r, by definition. The polygon P is thus dissected into two rectangles, each having B_1C_1 as one side, and n kite-shaped quadrilaterals which as before can be fitted together to form an n-gon denoted by P^*. Also as before, P^* is a polygon circumscribing a circle of radius r. Corresponding to equations (4) we have in this case the results

$$A = 2rd + A^*, \quad L = 2d + L^*. \tag{5}$$

Equations (4) and (5) play an important role in the next two sections.

12.3. The Isoperimetric Theorem.

Theorem 12.3a. *Among all simple closed curves (in the plane) of a given perimeter, the circle encloses the largest area.*

A limited proof of this result was given in Theorem 4.3b, where the existence of a solution was assumed.

Another way to state this theorem is that the isoperimetric quotient is less than 1 for any simple closed curve other than a circle. The proof is separated into four parts.

Case 1. *Circumscribable polygons.* If P is a polygon circumscribing a circle of radius r, we prove that

$$2A = rL, \quad L^2 > 4\pi A, \quad \text{and} \quad L > 2\pi r, \tag{1}$$

where A is the area, and L the perimeter, of P. The equation $2A = rL$ can be derived by calculating the areas of the triangles into which P is separated by line segments drawn from the center of the circle to the vertices of P, using the area formula "one-half altitude times base."

Since the polygon P circumscribes a circle of radius r, it follows that $A > \pi r^2$, and

$$rL = 2A > 2\pi r^2, \quad L > 2\pi r, \quad \text{and} \quad 4\pi A = 2\pi rL < L^2.$$

Case 2. *Convex polygons.* We prove that $L^2 > 4\pi A$ holds for any convex n-gon, by use of the theory of the inner parallel polygon of the preceding section. The proof is by induction on n. If $n = 3$, the polygon is a triangle, and it is circumscribed about its own inscribed circle, and hence we conclude that $L^2 > 4\pi A$ by Case 1 above.

Assume the inequality for any convex polygon with fewer than n sides, and consider a convex n-gon P of area A and perimeter L. If P is a circumscribable polygon, we apply Case 1 above. Otherwise we use equations (4) or (5) from the preceding section, depending on whether the inner parallel polygon P_1 exists or degenerates into a line segment. If the inner parallel polygon P_1 exists, it has fewer sides than P, that is, fewer than n sides. Hence $L_1^2 > 4\pi A_1$ by the induction hypothesis. The polygon P^* is circumscribed about a circle, and consequently $(L^*)^2 > 4\pi A^*$ by Case 1 above. Also we have $L^* > 2\pi r$ by inequality (1) in Case 1, and hence $2L_1L^* > 4\pi r L_1$. Adding these inequalities we have, by equations (4) of the preceding section,

$$L^2 = L_1^2 + 2L_1L^* + (L^*)^2 > 4\pi A_1 + 4\pi r L_1 + 4\pi A^* = 4\pi A.$$

On the other hand, if the inner parallel polygon of P degenerates into a line segment, we use equations (5) from the preceding section. Again we have the inequalities $(L^*)^2 > 4\pi A^*$ and $L^* > 2\pi r$ from Case 1 above, because P^* is a polygon circumscribed about a circle of radius r. We conclude that

$$\begin{aligned} L^2 = (L^* + 2d)^2 &= (L^*)^2 + 4dL^* + 4d^2 \\ &> 4\pi A^* + 8\pi r d + 4d^2 \\ &= 4\pi\,(A^* + 2rd) + 4d^2 \\ &= 4\pi A + 4d^2 > 4\pi A. \end{aligned}$$

Case 3. *Nonconvex polygons.* The convex hull of a nonconvex polygon P, as illustrated in Figure 4.2a, is a convex polygon with a larger area, a smaller perimeter, and hence a larger isoperimetric quotient. Using Case 2 above, we conclude that the isoperimetric quotient of P is less than 1.

Case 4. *Simple closed curves other than polygons.* Suppose, contrary to what we want to prove, there is a simple closed curve, not a circle, such that $L^2 \le 4\pi A$. In case $L^2 = 4\pi A$, we use Theorem 4.3b to conclude that there is another simple closed curve of equal perimeter, but enclosing a larger area. Hence we may assume that we have a simple closed curve C such that $L^2 < 4\pi A$. Furthermore, we may assume that C is convex, since otherwise we can replace it by its convex hull as we did in Case 3 above, giving an even larger isoperimetric quotient.

Define $\beta = 4\pi A - L^2$, so that $\beta > 0$. Choose n points on C to form an approximating n-gon, with area denoted by A_2 and perimeter by L_2, as illustrated in Figure 12.3a. We may choose n sufficiently large so that A_2 is as close to A as we please, in particular

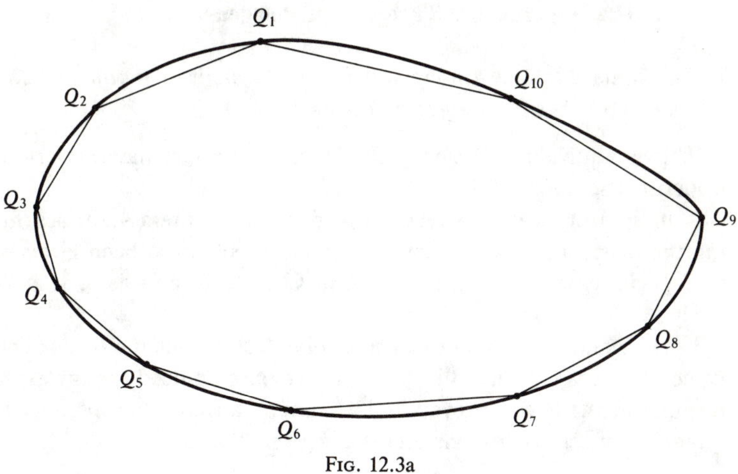

FIG. 12.3a

so that $A - A_2 < \beta/4\pi$. Also note that $L_2 < L$, because the straight-line path is shorter than the curve. Thus we can write

$$4\pi A_2 > 4\pi A - \beta = L^2 > L_2^2.$$

But the isoperimetric quotient of any polygon is less than 1 by the previous cases, and hence we have a contradiction. This completes the proof of the isoperimetric theorem.

The *Dido theorem* is an immediate consequence, that the semicircle encloses a greater area against a straight line than any other curve of the same length. (This was proved in Theorem 4.3a, assuming the existence of an optimal curve.) The result follows at once by using the mirror image in the straight line.

L1. Describe the shortest path across an equilateral triangle to bisect the area. *Suggestion*: Use the reflection principle. If the path cuts across an equilateral triangle ABC from some point on AB to some point on AC, draw the mirror image ABC' of the triangle ABC using AB as the line of reflection, then the mirror image of the triangle ABC' in the line AC', and so on until a regular hexagon has been created with center A. The closed path encircling A bisects the area of this hexagon.

12.4. The Isoperimetric Theorem for Polygons.

THEOREM 12.4a. *Among all n-gons of a given perimeter, the regular n-gon has the largest area.*

This is equivalent to saying that a regular n-gon has the largest isoperimetric quotient $4\pi A/L^2$.

A limited proof of this result was given in Theorem 4.2b, assuming the existence of a solution. Complete proofs have been given in the special cases $n = 3$, 4, and 6, in §3.2, §3.3, and §3.5, respectively.

Let c_n denote the isoperimetric quotient of a regular n-gon. (The value of c_n was calculated in §6.1, but we have no need for any exact formula here.) It was proved in Theorem 6.2a that the isoperimetric quotients of regular n-gons increase with n, that is,

$$c_3 < c_4 < c_5 < c_6 < \cdots. \tag{1}$$

Now we turn to the proof of Theorem 12.4a, separating the argument into three cases.

Case 1. *Circumscribable polygons.* It was proved in Theorem 6.3c that among all n-gons circumscribed about a circle, the regular n-gon has the largest isoperimetric quotient.

Case 2. *Convex polygons in general.* In view of Case 1, we may restrict attention to convex polygons that do not circumscribe any circle. The proof is by induction on n, the number of sides of the polygon. For $n = 3$, the result was proved in Theorem 3.2a. This provides a basis for the induction, and we assume that the result holds for all convex polygons with fewer than n sides. If P is a convex n-gon with area A and perimeter L, we prove that $4\pi A < c_n L^2$.

To do this, we apply the construction of §12.2 to the polygon P. Consider first the case where, as illustrated in Figure 12.2c, there is an inner parallel polygon P_1, with area denoted by A_1, and perimeter by L_1. Equations (4) of that section apply in this case, so that $4\pi A < c_n L^2$ can be written as

$$4\pi(A_1 + rL_1 + A^*) < c_n(L_1 + L^*)^2.$$

We prove this by establishing the three inequalities

$$4\pi A_1 < c_n L_1^2, \quad 4\pi r L_1 \le 2c_n L_1 L^*, \quad \text{and} \quad 4\pi A^* \le c_n(L^*)^2, \tag{2}$$

the first of which is a strict inequality. Since the inner parallel polygon P_1 has fewer than n sides, we obtain $4\pi A_1 \le c_{n-1}L_1^2$ by the induction hypothesis, and the first of the inequalities (2) follows by the use of (1).

The third of the inequalities (2) is a consequence of Case 1, because P^* is an n-gon circumscribed about a circle of radius r. (It is possible that $4\pi A^* = c_n(L^*)^2$, because P^* might be a regular polygon.) Also we know that $2A^* = rL^*$ holds for P^*, by the first of formulas (1) in the preceding section. Substituting $A^* = \frac{1}{2}rL^*$ in $4\pi A^* \le c_n (L^*)^2$, we conclude that

$$2\pi r \le c_n L^*, \tag{3}$$

and the second inequality in (2) follows from this.

Finally, we turn to the case where the inner parallel polygon degenerates into a line segment of positive length d. (The case where the inner parallel polygon degenerates to a point occurs when the original polygon P is circumscribable, and this has been treated in Case 1 above.) Equations (5) of §12.2 are now valid, and we see that $4\pi A < c_n L^2$ can be written as

$$4\pi(2rd + A^*) < c_n(2d + L^*)^2.$$

To prove this, we establish the three separate inequalities

$$0 < 4c_n d^2, \quad 8\pi rd \le 4c_n dL^*, \quad \text{and} \quad 4\pi A^* \le c_n(L^*)^2, \tag{4}$$

which can be added to give the desired result. The first inequality here is obvious, the second is an immediate consequence of (3), and the third is the same as the third inequality in (2), already proved.

Case 3. *Nonconvex n-gons.* Given any nonconvex n-gon P with area A and perimeter L, we prove that $4\pi A < c_n L^2$. Let H be the convex hull of P, having area A_2 and perimeter L_2, say. Then $A_2 > A$ and $L_2 < L$. Furthermore, the convex polygon H has fewer than n sides, so that $4\pi A_2 < c_n L_2^2$ is a consequence of Cases 1 and 2 above and inequality (1). The inequality $4\pi A < c_n L^2$ follows at once, and the proof of Theorem 12.4a is complete.

12.5. Polygons with Prescribed Sides.

THEOREM 12.5a. *A polygon inscribed in a circle has larger area than any other polygon with sides of the same lengths in the same order.*

For quadrilaterals, this was proved in Theorem 3.3b. To establish the result in the general case, let P be a polygon inscribed in a circle of radius r, as illustrated in Figure 12.5a in the case of a quadrilateral. Let P_1 be a polygon not congruent to P, with sides of the same lengths in the same order. On each side of P_1 we construct an arc of a circle of radius r, giving segments of circles that are congruent to those surrounding the polygon P. This is illustrated in Figure 12.5b. Thus P_1 is surrounded by circular arcs of total length $2\pi r$. By the isoperimetric theorem, the area of P_1 and the circular segments enclosing it is less than the area of a circle of radius r. In the example, this means that the total area in Figure 12.5b is less than that in Figure 12.5a. Subtracting the circular segments, we conclude that the area of P_1 is less than that of P.

Theorem 12.5a is not true if the words "in the same order" are

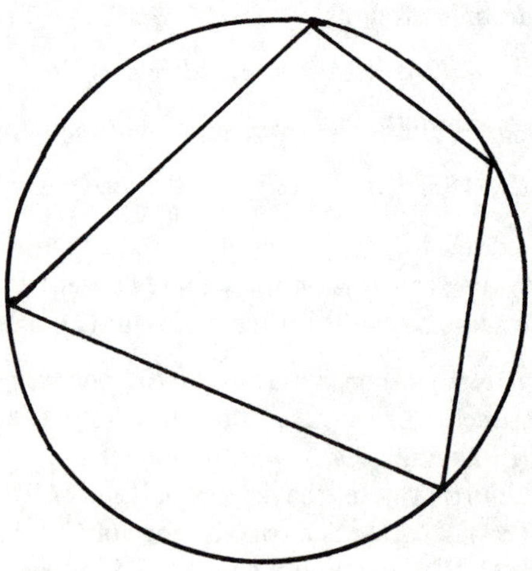

FIG. 12.5a

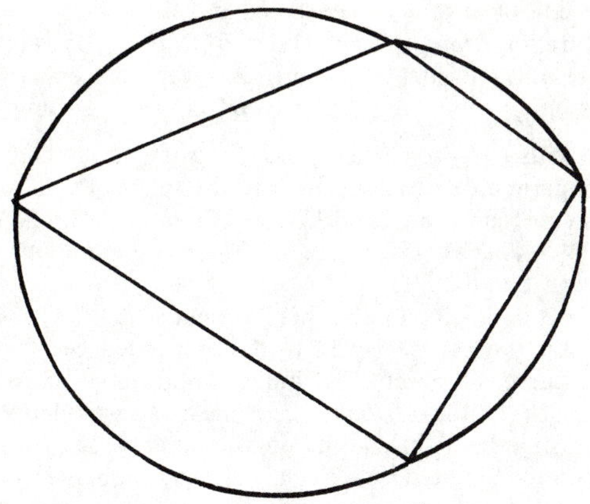

FIG. 12.5b

omitted. The reason for this is that, given any n-gon P inscribed in a circle of radius r, the circle can be separated into n sectors by joining the center to the vertices of P. These n circular sectors can be rearranged in any order we wish to form a circle of radius r having an inscribed n-gon with sides equal in length to those of P, but in a rearranged order.

Finally, we remark that Theorem 12.5a could have been proved as a consequence of the quadrilateral case in Theorem 3.3b, if we assume that there is a polygon P^* of largest area having sides of specified lengths. The argument is simply this: By Theorem 3.3b, every four consecutive vertices of P^* lie on a circle, and hence all vertices lie on a circle.

Notes on Chapter 12

§12.1. For other discussions of the significance of the existence question in isoperimetric results, see Courant and Robbins (1947, pp. 370–376), Kazarinoff (1961, pp. 58–63), and Pólya (1954, Ex. 1, pp. 181, 266).

§12.2. The origin of the inner parallel polygon is attributed by Fejes Tóth (1972, p. 27) to F. Riesz (1930). Riesz used the idea in quite a different mathematical setting.

§12.3. The line of argument developed in this section is a considerably modified adaptation from Benson (1966, pp. 127, 128) and Fejes Tóth (1972, pp. 8–11). In particular, where these authors employ a multiple-positioned version of the inner parallel polygon, we use a single position.

In contrast with our geometric approach, analytic methods provide approaches to isoperimetric problems beyond the scope of this book. Classical statements may be found in Bliss (1925) and Courant (1940, *Calculus*, vol. 2, pp. 213–214, 493–494, 518–520). See also the recent comprehensive studies by Osserman (1978 and 1979).

The arguments in §12.3 and elsewhere are not strictly "without calculus," because the limit process is needed to define the length of, and the area enclosed by, a curve. However, any simple closed *convex* curve does have a well-defined length (i.e., is *rectifiable*) and encloses a well-defined area. This can be proved using the definitions of length and area, and the easily proved result that a bounded increasing sequence of real numbers has a limit. As to nonconvex simple closed curves, we pass at once to the convex hull of any such curve, with an increase in area and a decrease in perimeter. Thus, although our discussions are based on intuition to some extent, a more rigorous formulation is not very esoteric.

POSTSCRIPT ON CALCULUS

Most of the questions discussed in this book are more manageable by the methods given here than by calculus. This is hardly surprising, since the problems have been selected to meet that general specification. Conversely, there are extremal problems in calculus books that are not solvable, not in any neat way, by the methods used here. Differential calculus can deal with a wider variety of functions.

However, in many problems there is no simple function to be subjected to the differentiation process. One example of this is the question of finding the quadrilateral of largest area among those having a specified perimeter. To describe such a quadrilateral in terms of sides and angles, four variables are needed, to denote for example the lengths of three sides and the size of one angle. The formulation of the area function in terms of four such variables is not an auspicious way to start solving the problem.

Another example is the problem of locating five points on the circumference of the unit circle so that the sum of the ten distances linking all pairs of distinct points is a maximum. (This is a paraphrase of Problem B1 of the William Lowell Putnam Mathematical Competition for 1974.) Key words like "maximum" and "minimum" are so strongly associated with calculus in the minds of some students that they immediately seek a function whose derivative will lead to a speedy solution. They sometimes plow right ahead with a function, usually a multivariable function, whose derivatives are hopelessly complicated.

The problem above of maximizing the sum of the ten distances is a case in point. Five points on the unit circle determine five angles at the center, subtended by consecutive pairs of the points. If these five angles, with sum 2π, are denoted by $2\theta_1, 2\theta_2, \ldots, 2\theta_5$, then the sum to be maximized is

$$\sum_{i=1}^{5} \sin \theta_i + \sum_{i=1}^{5} \sin(\theta_i + \theta_{i+1}),$$

where the identification $\theta_6 = \theta_1$ is needed in the second sum. By applying Theorem 5.2b to each of these sums separately, we are led at once to equal angles $\theta_1, \theta_2, \ldots, \theta_5$, which leads to the solution of the problem. By way of contrast with this solution, taking the derivatives of these sums leads to a messy situation.

These observations are not intended as any disparagement of calculus, which is a powerful instrument in the mathematical toolchest. Our point is simply that it is not the *only* instrument.

We mentioned above that calculus can handle a wider variety of functions than the methods of this book. Striking illustrations of this are the exponential and logarithmic functions. Since calculus provides the best mechanism for even defining these functions, it is to be expected that it also provides the techniques for working with them. Hence it is no accident that, except for our brief discussion here, the exponential and logarithmic functions are virtually absent from this book.

The exponential functions provide a very easy way of proving one of our central results, the inequality of the arithmetic-geometric means. We are including such a proof here at the suggestion of Gerald L. Alexanderson, who reminded us of the very elegant proof on page 103 of the book by Hardy, Littlewood, and Pólya (1952). Here is an adaptation of that proof.

The only background needed is the following simple proposition about the exponential function.

THEOREM 1. *If x is any real number, then $e^x \geq ex$, with equality iff $x = 1$.*

This follows at once from the observation that the function $f(x) = e^x - ex$ has an absolute minimum at $x = 1$, because the derivative $f'(x) = e^x - e$ vanishes iff $x = 1$, and the second derivative $f''(x) = e^x$ is positive for all values of x.

Now consider any nonnegative real numbers $a_1, a_2, \ldots, a_n$ with arithmetic mean A and geometric mean G. To prove that $A \geq G$, we substitute $x = a_1/A$, $x = a_2/A$, ..., $x = a_n/A$, successively into $e^x \geq ex$, and multiply the inequalities so obtained. Since $a_1 + a_2 + \cdots + a_n = nA$, the outcome is

$$e^n \geq e^n(a_1/A)(a_2/A) \cdots (a_n/A) = e^n G^n/A^n.$$

Canceling the factor e^n, this gives $1 \geq G^n/A^n$, so that $A \geq G$. Furthermore, there is equality in $A \geq G$ iff each of the substituted values for x is 1; that is, $a_1/A = 1$, $a_2/A = 1$, ..., $a_n/A = 1$. This leads at once to the condition $a_1 = a_2 = \cdots = a_n$.

But the real power of calculus can be demonstrated here by the observation that the inequality $A \geq G$ can be obtained just as easily in the far more general form with *weighted* means, as follows.

THEOREM 2. *Let $q_1, q_2, \ldots, q_n$ be positive real numbers with sum* 1. *Then, for any nonnegative real numbers $a_1, a_2, \ldots, a_n$, we have*

$$q_1a_1 + q_2a_2 + \cdots + q_na_n \geq a_1^{q_1}a_2^{q_2} \cdots a_n^{q_n}, \qquad (1)$$

with equality iff $a_1 = a_2 = \cdots = a_n$.

The arithmetic-geometric inequality is the special case of this with each of $q_1, q_2, \ldots, q_n$ assigned the value $1/n$.

To prove the theorem we denote the weighted sum on the left side of (1) by w, and thus $w = q_1a_1 + \cdots + q_na_n$. We then replace x in Theorem 1 by a_j/w to get

$$e^{a_j/w} \geq ea_j/w \quad \text{or} \quad e^{a_jq_j/w} \geq (ea_j/w)^{q_j},$$

by raising the inequality to the power q_j. Multiplying these inequalities for each of the values $j = 1, 2, \ldots, n$, and using the definition of w and the fact that $q_1 + q_2 + \cdots + q_n = 1$, we get

$$e \geq e\,[a_1^{q_1}a_2^{q_2} \cdots a_n^{q_n}]/w,$$

which establishes (1). Moreover, there is equality here iff each of a_1/w, a_2/w, ..., a_n/w equals 1, giving the usual condition $a_1 = a_2 = \cdots = a_n$.

There are two other immediate consequences of Theorem 1 that we note.

THEOREM 3. *The maximum value of $x^{1/x}$ over all positive real numbers x is $e^{1/e}$.*

To see this, we replace x by x/e in the inequality of Theorem 1 to get

$$e^{x/e} \geq x \quad \text{or} \quad e^{1/e} \geq x^{1/x},$$

with equality iff $x/e = 1$ or $x = e$.

THEOREM 4. *If* (i) $b > a \geq e$, *or if* (ii) $0 < b < a \leq e$, *then* $a^b > b^a$.

A special case of this is the well-known result $e^\pi > \pi^e$.

To prove the theorem we start with case (i) and observe that $a/e \geq 1$ and $b - a > 0$, and hence

$$(a/e)^{b-a} \geq 1 \quad \text{or} \quad a^b e^a \geq e^b a^a. \tag{2}$$

Replacing x by b/a in Theorem 1 we get

$$e^{b/a} > eb/a \quad \text{or} \quad e^b a^a > e^a b^a, \tag{3}$$

with strict inequality because $b/a \neq 1$. Results (2) and (3) imply $a^b e^a > e^a b^a$, giving the result in the theorem.

In case (ii) we have $e/a \geq 1$ and $a - b > 0$, so that

$$(e/a)^{a-b} \geq 1 \quad \text{or} \quad a^b e^a \geq e^b a^a,$$

which is the same as the second inequality in (2) above. The rest of the proof is the same as in case (i).

M1. With or without calculus, find the largest real number x such that the sequence

$$x,\ x^x,\ x^{x^x},\ x^{x^{x^x}},\ \ldots$$

converges to a finite limit. What is the limit?

Notes

The position taken in this book is that, while calculus offers a powerful technique for solving extremal problems, there are other methods of great power that should not be overlooked. A typical contrasting view is that expressed by C. Stanley Ogilvy (1968, p. 35), quoted here with the permission of the publisher, Prindle, Weber, and Schmidt:

> The basic methods for finding maxima and minima of nonlinear functions are analytic: the familiar calculus method and an extension of it using

> Lagrange multipliers. We have already suggested that occasionally special methods would yield answers to extremal problems, without any calculus. These are "trick methods," each applying solely to its own problem. Usually they cannot be extended, lacking the great generality and power of the analytic (calculus) methods. Their chief interest lies in their novelty.

Curiously enough, on p. 37 of this account Ogilvy writes, "It is well known that of all polygons of n sides and constant perimeter, the regular n-gon has the largest area," with no proof given. This important result is a good illustration of a problem best solved by methods other than calculus.

Many students, following the viewpoint expressed above, try to solve the extremal questions on the Putnam exams by seeking some function to differentiate, even though most of these problems are handled best by other methods. Moreover, students will often pursue the differentiation process through thick and thin, in spite of the hopelessly complicated functions at hand.

The inequality in Theorem 4 has been rediscovered many times. For other approaches see Just and Schaumberger (1975), Niven (1972), and Varner (1976).

The book *Selected Papers on Calculus,* edited by Tom M. Apostol et al. (1968), contains in Chapter 8 a wealth of material on maxima and minima. In addition to the 17 papers reprinted in this chapter, there is a bibliography on pp. 270–271.

SOLUTIONS OF PROBLEMS

A1. Use formula (3) with $\theta + \lambda = 180°$, so that $\cos(\theta + \lambda) = -1$. Then prove $s - a = b$ and $s - c = d$.

A2. Yes. The proof of (3) is valid exactly as given if we take x to be the length of the interior diagonal.

A3. Taking an arbitrarily selected unit of length to start with, let the area be $A = \frac{1}{2}bh$, where b and h are the lengths of the base and the altitude. Then if the unit of length is redefined to be $\sqrt{A}$ times as long as the original unit, the base and the altitude have lengths $b/\sqrt{A}$ and $h/\sqrt{A}$ when measured in the new unit. Hence the area is now 1.

A4. The case $a = 3$, $b = 4$, $c = 6$ shows that $a + b > c$ does not imply $a^2 + b^2 > c^2$. However, it does imply $\sqrt{a} + \sqrt{b} > \sqrt{c}$, because if $\sqrt{a} + \sqrt{b} \leq \sqrt{c}$ then by squaring we get $a + 2\sqrt{ab} + b \leq c$, and this contradicts $a + b > c$.

A5. The conclusion $a > d$ is valid, for example, by the argument

$$a > (a + b)/2 > (c + d)/2 > d.$$

The other three, $a > c$, $b > c$, and $b > d$, do not necessarily follow, as the examples $a = 10$, $b = 2$, $c = 6$, $d = 5$ and $a = 10$, $b = 8$, $c = 11$, $d = 5$ show.

A6. If $a > 1$, then $\max(a, a^2) = a^2$, whereas if $0 < a < 1$, we have $\max(a, a^2) = a$. In the second part of the problem the maximum is $\sqrt{a} + \sqrt{b}$.

A7. Inequality (ii).

A8. Yes. This follows from the triangle inequality and the fact that sin α, sin β, and sin γ are proportional to the lengths of the sides.

A9. Using $a + b > 2\sqrt{ab}$ with a and b replaced by x_1 and c/x_1 we see that

$$x_2 = \frac{1}{2}(x_1 + c/x_1) > \sqrt{c}.$$

It follows that $x_n > \sqrt{c}$ if $n > 1$. Next $x_2 > \sqrt{c}$ implies that $c/x_2 < \sqrt{c}$. Adding x_2 and using the definition of x_3 we get $2x_3 < x_2 + \sqrt{c}$, and this is equivalent to $x_3 - \sqrt{c} < \frac{1}{2}(x_2 - \sqrt{c})$. The same argument holds with x_2 and x_3 replaced by x_n and x_{n+1}.

A10. No. In case $A = \{9, 4, 2\}$, $B = \{8, 6, 1\}$, and $C = \{7, 5, 3\}$ we see that $A > B$, $B > C$, but $C > A$.

B1. Writing $u = ax$, $v = by$, we see that the product uv is constant because $uv = abxy = abc$. By Corollary 3 the minimum value of $u + v$ is obtained by taking $u = v$; thus $u = v = \sqrt{abc}$, and the minimum is $2\sqrt{abc}$. But $u + v = ax + by$; so the minimum value of $ax + by$ is also $2\sqrt{abc}$, obtained by taking

$$ax = by = u = v = \sqrt{abc}, \quad x = \sqrt{bc/a}, \quad y = \sqrt{ac/b}.$$

Note that $xy = c$, as required.

B2. By Theorem 2.2b we know that

$$a^2 + b^2 \geq 2ab, \quad a^2 + c^2 \geq 2ac, \quad b^2 + c^2 \geq 2bc.$$

Now a, b, and c are not all equal, so at least one of these inequalities is strict; for example, if $b \neq c$ then $b^2 + c^2 > 2bc$. Hence if we add the inequalities we get the strict inequality

$$2a^2 + 2b^2 + 2c^2 > 2ab + 2ac + 2bc.$$

B3. The number is $\frac{1}{2}$, by the following argument. If x is the number in question, it exceeds its square by $x - x^2$. By Theorem 2.2a the maximum of $cx - x^2$ occurs if $x = c/2$, so the maximum of $x - x^2$ occurs if $x = \frac{1}{2}$.

B4. To prove that the harmonic mean can be written in the form $2ab/(a + b)$ is simple by elementary algebra. The inequalities

$$\sqrt{ab} > \frac{2\,ab}{a+b} \quad \text{and} \quad a+b > 2\sqrt{ab}$$

are equivalent, because the second can be obtained from the first by multiplying both sides by $(a + b)/\sqrt{ab}$.

B5. Write $u = x^4$ and $v = 2y^4$ and note that uv is a constant, $uv = 2x^4y^4 = 2c^4$. Hence $u + v$ is a minimum if $u = v = \sqrt{2}c^2$, by Corollary 3. The minimum value of $x^4 + 2y^4$ is therefore $2\sqrt{2}c^2$, obtained by taking $x = 2^{1/8}\,c^{1/2}$ and $y = 2^{-1/8}\,c^{1/2}$.

B6. The minimum of $x^2 + ax$ is $-a^2/2$ by taking $x = -a/2$, and the minimum of $y^2 + by$ is $-b^2/2$ with $y = -b/2$. Hence the minimum asked for is $c - \frac{1}{2}a^2 - \frac{1}{2}b^2$.

B7. These results follow from the identities

$$2cx - x^2 = c^2 - (x - c)^2, \quad x + c^2/x = (\sqrt{x} - c/\sqrt{x})^2 + 2c.$$

B8. Adding $xr \leq (x^2 + r^2)/2$ and $ys \leq (y^2 + s^2)/2$ we get $xr + ys \leq (x^2 + y^2 + r^2 + s^2)/2 = 1$. Thus the maximum is 1 because it is attained, for example, with $x = r = 1$, $y = s = 0$.

B9. 405, obtained by taking $20x = y$.

B10. $30\sqrt{3}$, or 51.96 to two decimal places. At a speed of x miles per hour, the trip requires $400/x$ hours. The driver's wages are $3200k/x$ dollars. The fuel cost per hour is $k(1 + x/40 + x^2/300)$ dollars, and hence the total fuel cost is obtained by multiplying this by $400/x$ to give $k(400/x + 10 + 4x/3)$. Adding $3200k/x$ we see that the total cost of the trip is $k(3600/x + 10 + 4x/3)$. Ignoring the constants k and 10, we want to minimize $3600/x + 4x/3$. These two terms have a constant product, and consequently we get $3600/x = 4x/3$, as in Example 2 in the text.

B11. With x and y replaced by G and xy/G, we must prove that $x + y > G + xy/G$. Multiplying by G and rearranging terms, this amounts to

$$(G - x)(y - G) > 0.$$

This is correct because the geometric mean G of a set of numbers not all equal lies between the smallest number x and the largest number

y; thus $x < G < y$. The process replaces $a_1 + a_2 + \cdots + a_n$ by $G + G + \cdots + G$, because at least one more term G is introduced into the sum at each step.

B12(i). 125/27. Since $x + y + z$ is a constant we take $x = y = z = 5/3$; so the answer is $(5/3)^3$.

(ii). 72. The factors of the product $(2x)(3y)(4z)$ have a constant sum, so the maximum value of the product is obtained by taking $2x = 3y = 4z = 12$, giving $x = 6$, $y = 4$, $z = 3$. Also, in maximizing $(2x)(3y)(4z)$ we are simultaneously maximizing xyz; so we have $xyz = 72$.

B13. $k^3/27abc$. As in part (ii) of the preceding problem, we start by maximizing $(ax)(by)(cz)$, giving $ax = by = cz = k/3$.

B14. 75. Since the factors of the product $(3x)(3x)(5y)$ have a constant sum, this product is maximized by taking $3x = 3x = 5y = 15$, giving $x = 5$, $y = 3$.

B15. 6. Set $x = y + z$, so that $z > 0$. The problem then is to minimize $y + z + 8/yz$. The product $yz(8/yz)$ is constant, so we solve $y = z = 8/yz$ to get $y = z = 2$ and $x = 4$.

B16. Taking reciprocals we seek the minimum of each of

$$x + \frac{a}{x}, \quad x + \frac{a}{x^2}, \quad x^2 + \frac{a}{x}.$$

The first of these is the sum of two terms with a constant product a, so we use $x = a/x$ or $x = \sqrt{a}$. The second can be written as

$$\frac{x}{2} + \frac{x}{2} + \frac{a}{x^2},$$

three terms with a constant product $a/4$, so we write $x/2 = a/x^2$, or $x = (2a)^{1/3}$. The third sum was treated in Example 3 in the text of §2.6. Thus we see that the maximum values asked for in the problem are

$$\frac{1}{2\sqrt{a}}, \quad \left(\frac{4}{27a}\right)^{1/3}, \quad \text{and} \quad \left(\frac{4}{27a^2}\right)^{1/3}.$$

B17. Squaring, we seek the maximum of $x^2(1 - x^2)$, a product of two terms with constant sum 1. Thus we have a maximum iff $x^2 = 1 - x^2$ or $x = \sqrt{1/2}$. Thus the maximum value asked for is ½.

B18. The given function is negative if $x > \sqrt{12}$, so we restrict attention to numbers x satisfying $0 < x \leq \sqrt{12}$. Squaring the function we obtain

$$2(2x^2)(12 - x^2)(12 - x^2).$$

Ignoring the initial multiplier 2 for a moment, we see that the three terms of the product have a constant sum 24. Hence we write $2x^2 = 12 - x^2$, $3x^2 = 12$, $x = 2$. Hence the solution of the problem is 32.

B19. Writing the sum as $r^2 + \frac{1}{2}rh + \frac{1}{2}rh$, we have a sum of three terms with product $r^4h^2/4$, and this equals the constant $c^2/4$. Hence we get a minimum by taking $r^2 = \frac{1}{2}rh$, so that $h = 2r$. This with $r^2h = c$ gives $2r^3 = c$ or $r = (c/2)^{1/3}$, and the minimum value requested is $3(c/2)^{2/3}$.

B20. Denoting the number by x, we want to maximize $x - x^3$ among positive numbers x. This is $x(1 - x^2)$ and, as in B3 above, we square this and write it as

$$\frac{1}{2}(2x^2)(1 - x^2)(1 - x^2).$$

Apart from the factor ½, we have a product of three terms with constant sum 2. Hence we write $2x^2 = 1 - x^2$, and so the answer is $x = \sqrt{1/3}$.

B21. The problem is to maximize $x^2 - x^3$, or $x^2(1 - x)$, among positive numbers x. Writing the product as

$$4(x/2)(x/2)(1 - x),$$

we see that the factors $x/2$, $x/2$, and $1 - x$ have a constant sum. Hence we write $x/2 = 1 - x$, giving the answer $x = 2/3$.

B22. The terms of the sum have constant product 1000, so we write $50/x = 20/y = xy$. The only solution in positive numbers is $x = 5$, $y = 2$, so the answer to the question is 30.

B23. Ignoring the constant 21 we note that the fractions x/y, $2y/z$, $4z/x$ have a constant product 8. So we want to solve

$$\frac{x}{y} = \frac{2y}{z} = \frac{4z}{x}.$$

This gives $x = 2y = 2z$, so that only the ratios between x, y, z can be determined. The minimum value asked for is 18.

B24. Write the sum as $3x + 3x + 24/x^2$ so that the three terms have constant product 216. So the minimum is 18, attained when $x = 2$.

B25. 1/256. Writing the expression as

$$\left(1 - \frac{2}{y} - \frac{2}{y} - \frac{3}{x}\right)\left(\frac{2}{y}\right)\left(\frac{2}{y}\right)\left(\frac{3}{x}\right),$$

we see that the factors have a constant sum 1; so we set each factor equal to ¼, with $x = 12$ and $y = 8$.

B26. The product of the terms in the sum to be minimized is (xy) $(2xz)$ $(3yz) = 6x^2y^2z^2$, a constant. Hence the minimum is found by taking $xy = 2xz = 3yz$ if these equations can be solved with the given condition $xyz = 48$. There is a unique solution in positive numbers, $x = 6$, $y = 4$, $z = 2$; so the minimum value is 72.

B27. The product of the three terms x^2, $12y$, and $10xy^2$ in the sum to be minimized is a constant, but then it turns out that the equations $x^2 = 12y = 10xy^2$ and $xy = 6$ have no solutions. So we try another method, and use the substitution $y = 6/x$ in the sum:

$$x^2 + 12y + 10xy^2 = x^2 + 72/x + 360/x = x^2 + 432/x.$$

If we write this sum as $x^2 + 216/x + 216/x$, we see that the terms x^2, $216/x$, and $216/x$ have a constant product; so we write $x^2 = 216/x$. This leads to $x = 6$, and $y = 1$, giving the answer 108.

B28. We multiply out the product to get

$$3 + (x/y + y/x) + (x/z + z/x) + (y/z + z/y).$$

The minimum of each term in parentheses is 2, occurring iff $x = y = z$. Thus the minimum value is $3 + 2 + 2 + 2$ or 9. The minimum of $x^{-1} + y^{-1} + z^{-1}$ must be $9/c$, since $x + y + z = c$.

B29. If we add the inequalities $a_i^2 + a_j^2 \geq 2a_ia_j$ for all pairs of integers i and j satisfying $1 \leq i < j \leq n$, it is readily observed that each of $a_1^2, a_2^2, \ldots, a_n^2$ occurs $n - 1$ times in the sum on the left, because each of these squares appears once with each of the others. This gives the inequality asserted. Furthermore, if any two of the a's are unequal, say, for example, $a_2 \neq a_4$, then we know that $a_2^2 + a_4^2 > 2a_2a_4$ and so there is strict inequality in the entire sum.

B30. The root mean square is positive in all cases, unless all the numbers are zero, in which case $R = A = 0$. Otherwise it suffices to prove $R^2 \geq A^2$; that is,

$$(a_1^2 + a_2^2 + \cdots + a_n^2)/n \geq (a_1 + a_2 + \cdots + a_n)^2/n^2.$$

Multiplying both sides by n^2, and then squaring out the term on the right, we see that by subtracting $a_1^2 + a_2^2 + \cdots + a_n^2$ the inequality reduces to the result in problem B29.

B31. Squaring out all the terms like $(a_1 - x)^2$ in $f(x)$, we see that

$$\begin{aligned} f(x) - a_1^2 - a_2^2 - \cdots - a_n^2 &= nx^2 - 2x(a_1 + a_2 + \cdots + a_n) \\ &= n[x^2 - 2xA]. \end{aligned}$$

By Theorem 2.2a the minimum occurs if $x = A$.

B32. The numbers 1, 2, 2, 2, 2 will do, and 1, 1, 1, 1, 2 for the second example.

B33. This is clear from the definitions in equation (1), because

$$(1 + 1/n)^n < (1 + 1/n)^n\,(1 + 1/n) = (1 + 1/n)^{n+1}.$$

B34. This follows from $f(m) < g(m)$ and $g(m) < g(k)$, the first of which comes from B33 and the second from the sequence (3).

B35. The algebra needed here is

$$\begin{aligned} (1 + 1/n)^{n+1} - (1 + 1/n)^n &= (1 + 1/n)^n \cdot [(1 + 1/n) - 1] \\ &= (1 + 1/n)^n \cdot [1/n]. \end{aligned}$$

B36. $\sqrt[3]{3}$. Noting that $\sqrt{2} = \sqrt[4]{4}$, we prove by induction that $\sqrt[3]{3} > \sqrt[n]{n}$ for $n \geq 4$. The case $n = 4$ is easily checked. We assume $3^n > n^3$, and prove $3^{n+1} > (n+1)^3$. Multiplying by 3 we have $3^{n+1} > 3n^3$; so it suffices to prove $3n^3 > (n+1)^3$. This follows by adding $n^3 = n^3$, $n^3 > 3n^2$, and $n^3 > 3n + 1$.

B37. -8; 1. Apply Corollary 1 of Theorem 2.2a to $x^2 + 6x + 1$. As to $x^4 + 6x^2 + 1$, note that x^4 and x^2 are nonnegative.

B38. The maximum is $1/8$. Writing the expression in the form

$$(54)(108) \cdot \frac{x}{2} \cdot \frac{x}{2} \cdot \frac{y}{3} \cdot \frac{y}{3} \cdot \frac{y}{3}(1 - x - y),$$

we see that the sum of the factors is a constant. Thus we get a maximum by writing $x/2 = y/3 = 1 - x - y$. These equations have the solution $x = 1/3$, $y = 1/2$.

B39. The inequality $x^3 + 1 \geq 2x$ does not hold for all positive x, for example $x = 4/5$. The largest value of k is $3/\sqrt[3]{4}$. To prove this we use Theorem 2.3a to get

$$x^3 + 1 = x^3 + 1/2 + 1/2 \geq 3(x^3/4)^{1/3}$$

for all positive x, with equality iff $x^3 = 1/2$.

B40. Equality holds if $a^2 = b^3 = c^6$. The inequality can be proved by applying Theorem 2.5a in the form $A \geq G$ with $n = 6$ to the quantities $a^2, a^2, a^2, b^3, b^3, c^6$.

B41. From $(a + b)^2 = (c + d)^2$ and $a^2 + b^2 > c^2 + d^2$ it follows that $ab < cd$. The result follows if we add the inequalities $(a + b)(a^2 + b^2) > (c + d)(c^2 + d^2)$ and $-ab(a + b) > -cd(c + d)$.

B42. The largest value is 3; there is no smallest. Observe that the reciprocal of the given expression equals

$$[\sqrt{(x + 2)^2 + 81} + \sqrt{(x + 2)^2 + 36}\,]/45,$$

which has no maximum, but has its minimum at $x = -2$.

B43. $k = 62$. With $k = 1, 2, 3, \ldots, 60$, note that a solution for n can be found in each of these cases among $n = 9, 18, 27, \ldots, 990$.

Also, $n = 1098$ is a solution of $n = 61S(n)$. To prove that $n = 62S(n)$ has no solution, first observe that if n has 3 digits, say s, t, and u, then $9(11s + t) = 61(s + t + u)$, which is easily proved to have no solutions in appropriate digits. If n has 4 digits, say r, s, t, and u, then $9(111r + 11s + t) = 61(r + s + t + u)$, which again has no solutions in digits, by using the facts that 61 is a divisor of $111r + 11s + t$, and 9 is a divisor of $r + s + t + u$. If n has 5 digits, then $S(n) \leq 45$, giving $n = 62S(n) \leq 2790$, a contradiction. This argument holds a fortiori for integers n with more than 5 digits.

B44. $x = -b/2a$.

B45. The minimum value is 15. This can be obtained by applying the transformation $x = X - 2y$ to the given quadratic polynomial, reducing it to $4X^2 + 9y^2 - 24X + 18y + 60$. This is minimized by taking $X = 3$, $y = -1$, or $x = 5$, $y = -1$ in the original polynomial.

B46. There is no unique way of stating necessary and sufficient conditions in this problem. One formulation is that either the conditions $a > 0$ and $ac > b^2$ hold, or the conditions $a > 0$, $ac = b^2$, and $ae = bd$. (The inequality $a > 0$ can be replaced by $c > 0$, and $ae = bd$ can be replaced by $be = cd$, if we want an alternative statement.) To prove this, we note first that $a > 0$ and $c > 0$ are necessary conditions. For if $a < 0$ it is clear that $f(x, 0) = ax^2 + dx + k$ has no minimum: it is unbounded below. Also we use

$$f(X - by/a, y) = aX^2 + (c - b^2/a)y^2 + dX + (e - db/a)y + k.$$

If $c - b^2/a < 0$, there is no minimum. If $c = b^2/a = 0$, then f has a minimum iff the coefficient of y is zero, which leads to the condition $ae = bd$.

C1. This is just another way of stating Theorem 3.2a. For an equilateral triangle with sides of length a, the area A and the perimeter L are $\sqrt{3}a^2/4$ and $3a$, respectively, so that $A = \sqrt{3}L^2/36$. This becomes an inequality for any triangle that is not equilateral.

C2. The isosceles right triangle in both cases. Denoting the sides by x and y, the hypotenuse is $\sqrt{x^2 + y^2}$. To maximize the area we seek the largest value of $xy/2$. This amounts to maximizing x^2y^2, given that $x^2 + y^2$ is a constant. By Theorem 2.2a we take $x^2 = y^2$, or $x = y$. To maximize the perimeter, we look for the largest value of $x + y$, since the hypotenuse is fixed. We maximize $(x + y)^2$, that is, $2xy + x^2 + y^2$, or $2xy$ added to a constant. Again we get $x = y$.

C3. Using the suggestion in the problem, we see that

$$\begin{aligned} &PA^2 + PB^2 + PC^2 \\ &= (x - a_1)^2 + (y - a_2)^2 + (x - b_1)^2 + (y - b_2)^2 \\ &\quad + (x - c_1)^2 + (y - c_2)^2 \\ &= 3x^2 - 2x(a_1 + b_1 + c_1) + 3y^2 - 2y(a_2 + b_2 + c_2) \\ &\qquad + a_1^2 + a_2^2 + b_1^2 + b_2^2 + c_1^2 + c_2^2. \end{aligned}$$

We minimize the terms in x and y separately by use of Theorem 2.2a, Corollary 1.

C4. Write x, y, z for the sides PQ, PR, QR, respectively, so that $x + y = c$. The area of the triangle is $\frac{1}{2}xy \sin P$, and since the angle P is fixed, the problem in part (i) is to maximize xy. This gives $x = y$. In part (ii) the problem is to minimize z. By the law of cosines we have

$$z^2 = x^2 + y^2 - 2xy \cos P = c^2 - 2xy(1 + \cos P).$$

Therefore, to minimize z, we maximize xy, again getting $x = y$.

C5. The common area is 120. The circles are the same size, by the following argument. Let the polygon Q_1 with vertices P, R, S, T in that order be inscribed in a circle with center C. Then the sectors PCR, RCS, SCT, TCP can be reassembled in any order to form a circle of the same size with the quadrilateral Q_2 inscribed, or for that matter, with any quadrilateral with sides of lengths 8, 9, 12, 19 in any order we please.

C6. Apart from the side of specified length, the other three sides should be equal, and the quadrilateral should be inscribable in a circle. To prove this, let the sides have lengths a, x, y, z, where a is a constant. Write $s = (a + x + y + z)/2$, so that s is a constant. The inscribable quadrilateral then has area A, satisfying

$$A^2 = (s - a)(s - x)(s - y)(s - z).$$

The product $(s - x)(s - y)(s - z)$ is maximized by taking $x = y = z$.

C7. 18.

C8. Dimensions 150 by 300 feet, with the longer side parallel to the river. To prove this, let x be the length of a side perpendicular to the river, so the fencing is separated into lengths x, x, and $600 - 2x$. The problem is to maximize the area $x(600 - 2x)$ or $\frac{1}{2}(2x)(600 - 2x)$. Thus we write $2x = 600 - 2x$ since these two factors have a constant sum 600.

C9. The sides should have lengths 200, 200, 200, and 400 feet, the latter along the river, the angles should be 60°, 60°, 120°, 120° so that the quadrilateral is half of a regular hexagon. To prove this, let the sides be x, y, z, and w, the last one along the river. Since $x + y + z = 600$, the semiperimeter is $s = 300 + w/2$. The quadrilateral should be inscribable in a circle, so the area A satisfies $A^2 = (s - x)(s - y)(s - z)(s - w)$, and hence we have $3A^2$

$$= (300 - x + w/2)(300 - y + w/2)(300 - z + w/2)(900 - 3w/2).$$

These factors have a constant sum 1200; so we set each factor equal to 300. It follows that $x = y = z = w/2 = 200$.

C10. Consider any nonconvex quadrilateral $PQRS$, say with the vertex S lying inside the triangle PQR as shown in Figure C10, having the specified perimeter c. Write x, y, z, w for the lengths of the sides PQ, QR, RS, SP, where $x + y + z + w = c$. The area of the quadrilateral is less than the area of the triangle PQR. Since $PR < z + w$, the area of PQR is less than the area of a triangle with sides x, y, $z + w$. This triangle has perimeter c, and so by Theorem 3.2a its area is at most $c^2\sqrt{3}/36$, which is the area of an equilateral triangle of perimeter c. This proves the inequality in the problem.

To prove that there are nonconvex quadrilaterals of perimeter c with area as close to $c^2\sqrt{3}/36$ as we please, take $x = y = c/3$ and $z = w = c/6$, and take the point S just inside the triangle PQR. Thus S is almost on the line segment PR, whose length is almost s. The area of ΔPQR is almost $c^2\sqrt{3}/36$, in fact as close as we please to this value, and so is the area of the quadrilateral $PQRS$. The area $c^2\sqrt{3}/36$ cannot be attained by $PQRS$ because the point S must stay inside the triangle PQR for a nonconvex quadrilateral.

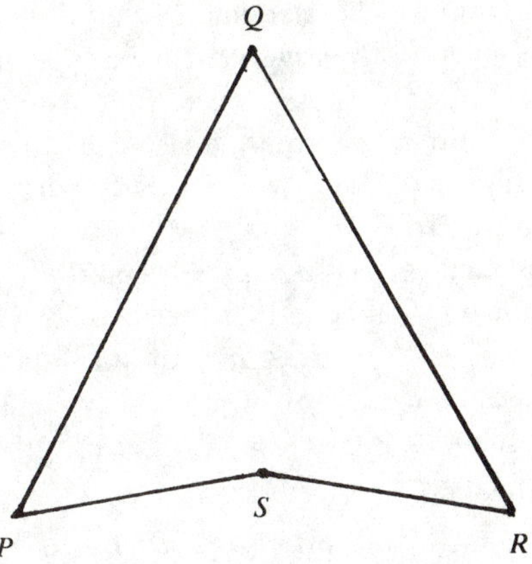

FIG. C10

C11. Let the vertices of the quadrilateral be P, Q, R, S in that order, and let the point of intersection of the diagonals be K. Then adding the inequalities $PK + KQ > PQ$ and $RK + KS > RS$, we find that the sum of the diagonals, $p_1 + p_2$, exceeds the sum of the lengths of a pair of opposite sides. Denote this result by $(*)$, for reference in the next problem. It follows that $a_1 + a_2 + a_3 + a_4 < 2p_1 + 2p_2$. Also $PQ + QS > QS$, and adding this to three analogous inequalities involving the sum of two adjacent sides we get $2(a_1 + a_2 + a_3 + a_4) > 2p_1 + 2p_2$.

C12. Of the eight inequalities of the form $p_i > a_j$, only two hold in all cases, namely, $p_2 > a_1$ and $p_2 > a_2$. Of the six inequalities of the form $p_1 + p_2 > a_r + a_s$, two fail in certain cases, namely, $p_1 + p_2 > a_2 + a_4$ and $p_1 + p_2 > a_3 + a_4$. The other four hold in all cases. Here is one way to prove these results. First we settle the failing cases by means of counterexamples. The rhombus with sides of length 1 and two opposite angles 30° shows the failure of $p_1 > a_j$ for $j = 1, 2, 3, 4$. The kite-shaped quadrilateral with vertices at $(\pm 5, 0)$, $(0, 10)$, and $(0, -1)$ on a rectangular coordinate system shows the failure of $p_2 > a_3$ and $p_2 > a_4$. The quadrilateral with vertices $(\pm 10, 0)$, $(\pm 1, 1)$ shows the failure of $p_1 + p_2 > a_2 + a_4$ and $p_1 +$

$p_2 > a_3 + a_4$. To prove the inequalities that do hold under all circumstances, we separate the argument for a moment into the three cases: that a_1 and a_2 are opposite sides, that a_1 and a_3 are opposite sides, and that a_1 and a_4 are opposite sides. By using the result $(*)$ from the solution in C11, we conclude that $p_1 + p_2 > a_3 + a_4$ holds in the first case, that $p_1 + p_2 > a_2 + a_4$ holds in the second case, and that $p_1 + p_2 > a_1 + a_4$ and $p_1 + p_2 > a_2 + a_3$ hold in the third case. It follows that these last two inequalities hold in all cases, since $a_1 \leq a_2 \leq a_3 \leq a_4$. An immediate consequence is that $p_1 + p_2 > a_1 + a_2$ and $p_1 + p_2 > a_1 + a_3$ hold in all cases. Finally, the inequality $p_1 + p_2 > a_2 + a_3$ implies $p_2 > a_2$ and $p_2 > a_1$ by simple logic.

C13. Choose the points so that angle $PCQ = 90°$. For if the circle has radius r, and the angle PCQ is denoted by θ in general, the area of the triangle is $\frac{1}{2}r^2 \sin \theta$. So the problem is to maximize $\sin \theta$.

C14. If the length, width, and height are denoted by x, y, z, then the volume is xyz and the surface area is $S = 2xy + 2yz + 2xz$. The product of the terms in this sum is fixed. So we minimize S by writing $2xy = 2yz = 2xz$, which leads to $x = y = z$.

C15. Using the dimensions x, y, z as in the preceding problem, we note that the volume equation is $xyz = k$. The problem now is to minimize the surface area with no top, that is $xy + 2xz + 2yz$. The terms in this sum have the product $4x^2y^2z^2 = 4k^2$. Hence we minimize the sum by taking $xy = 2xz = 2yz$, which gives $x = y = 2z$. These equations with $xyz = k$ yield the solution $x = y = \sqrt[3]{2k}$, $z = \sqrt[3]{k/4}$.

C16. $50 \times 50 \times 30$ feet, where 30 is the height. Let x, y, and z denote, respectively, the length, width, and height of the building, so that $xyz = 75{,}000$. Taking the heat loss through the floor as proportional to the area xy, the total loss of heat is proportional to

$$xy + 5xy + 10xz + 10yz \quad \text{or} \quad 6xy + 10xz + 10yz.$$

These three terms have a constant product; so a minimum is achieved by taking $6xy = 10xz = 10yz$.

C17. The largest possible volume is $abc/27$. Since the sum $x/a + y/b + z/c$ is a constant, the product of the terms $xyz/(abc)$ is a maximum, and so xyz is a maximum, if $x/z = y/b = z/c$. This leads to $x = a/3$, $y = b/3$, $z = c/3$; so $xyz = abc/27$.

C18. The dimensions for largest volume are $2a/\sqrt{3}$, $2b/\sqrt{3}$, $2c/\sqrt{3}$. The sum $x^2/a^2 + y^2/b^2 + z^2/c^2$ is constant; so the product of these three terms is a maximum if they are equal. This also makes xyz a maximum, and we get $x/a = y/b = z/c$, and the answer follows by simple algebra.

C19. The rectangle of largest area has dimensions $r\sqrt{2}$ by $r/\sqrt{2}$ with area r^2. To see this, take the vertices of the rectangle to be (x, y), $(x, 0)$, $(-x, y)$, and $(-x, 0)$, where x and y are positive and $x^2 + y^2 = r^2$. The dimensions of this rectangle are $2x$ and y; so the problem is to maximize $2xy$. Since r is a constant and $x^2 + y^2 = r^2$, we see that x^2y^2 is a maximum if $x^2 = y^2$ or $x = y$. The rest is simple algebra.

C20. The answer is 2000 cu. in. If the four square corners are of dimensions x by x, then the rectangular box has height x, and the bottom of the box has dimensions $30 - 2x$ by $30 - 2x$. The problem is to maximize the volume $x(30 - 2x)^2$. This can be written as

$$2(2x)(15 - x)(15 - x).$$

Ignoring the initial factor 2, the other three factors have the constant sum 30, so we set $2x = 15 - x$ to get $x = 5$.

C21. The volume is $\pi r^2 h$ and the surface area is $S = 2\pi r^2 + 2\pi rh$. Getting down to essentials by removing constant multipliers, the problem is to minimize $r^2 + rh$, given that r^2h is constant. We write $r^2 + rh$ in the form $r^2 + \frac{1}{2}rh + \frac{1}{2}rh$ so that the terms of this sum have a constant product,

$$(r^2)\left(\frac{1}{2}rh\right)\left(\frac{1}{2}rh\right) = \frac{1}{4}r^4h^2 = \frac{1}{4}(r^2h)^2.$$

Hence we set $r^2 = \frac{1}{2}rh$, giving $h = 2r$.

C22. The answer is $h = r$. The volume $\pi r^2 h$ is constant, and the surface area to be minimized is $\pi r^2 + 2\pi rh$ because the cylinder has no top. The problem is to minimize $r^2 + 2rh$, given that r^2h is constant. We write

$$r^2 + 2rh = r^2 + rh + rh$$

to get a sum of three terms whose product is constant. This leads to $r^2 = rh$ and so $r = h$.

C23. The relation is $h = r\sqrt{2}$. To prove this, we define the unit of length to be the radius of the given sphere. If A and D are the centers of the circular top and bottom of the inscribed cylinder, as shown in Figure C23, then the center C of the sphere is the midpoint of AD. Let B be any point on the rim of the top of the cylinder, so that B is also on the sphere. Thus we see that

$$BC = 1, \quad AB = r, \quad AC = h/2, \quad 1^2 = r^2 + (h/2)^2.$$

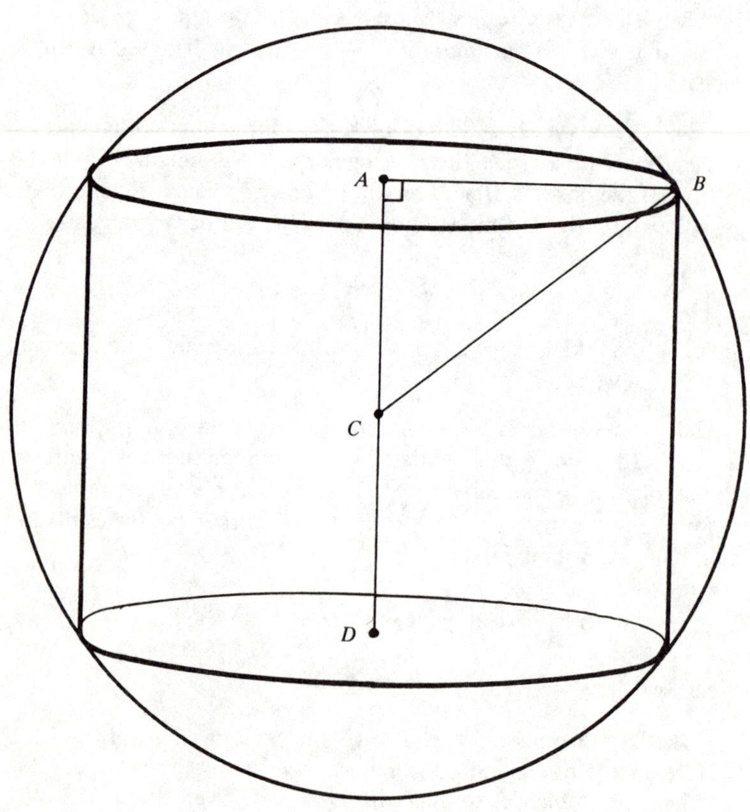

FIG. C23

The problem is to maximize the volume $\pi r^2 h$. To maximize $r^2 h$ we write the last equation as

$$2r^2 + 2r^2 + h^2 = 4,$$

so that the product of the terms in this constant sum is $(2r^2)(2r^2)(h^2) = 4r^4h^2$. Maximizing this will also maximize r^2h, so we write $2r^2 = h^2$, leading to the answer.

C24. Dimensions $b = 4\sqrt{3}$ and $h = 4\sqrt{6}$. The problem is illustrated in Figure C24, showing the right triangle with sides h, b, and 12. This gives $b^2 + h^2 = 144$, the problem being to maximize bh^2. We write

$$2b^2 + h^2 + h^2 = 288,$$

and the product of these terms is $2b^2h^4$. To maximize this, which also maximizes bh^2, we write $2b^2 = h^2$. Solving this equation with $b^2 + h^2 = 144$ gives the answer.

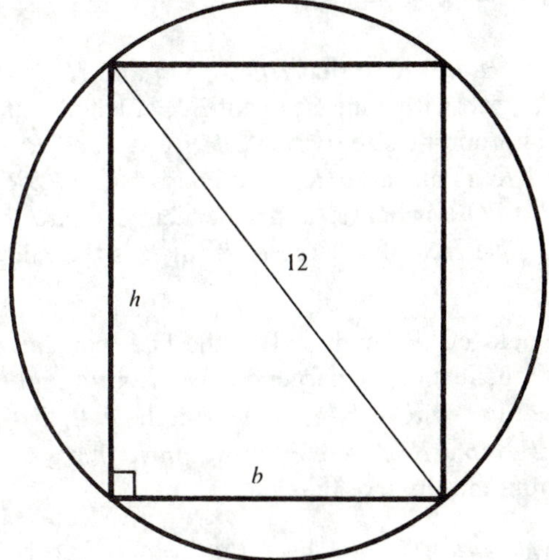

FIG. C24

C25. The square of the distance from $(0, c)$ to any point on the curve $y = x^2$ is

$$\begin{aligned}(x-0)^2 + (y-c)^2 &= x^2 + y^2 - 2cy + c^2 \\ &= x^4 - (2c-1)x^2 + c^2.\end{aligned}$$

Ignoring the constant c^2, we seek to minimize $x^4 - (2c - 1)x^2$. It is significant whether $2c - 1$ is positive, negative, or zero. If $c > 1/2$, so that $2c - 1$ is positive, we take $x^2 = (2c - 1)/2$ by §2.2. If $c = 1/2$ the problem is to minimize x^4; thus $x = 0$. If $c < 1/2$, so that $2c - 1$ is negative, then both terms x^4 and $-(2c - 1)x^2$ are positive for all values of x except $x = 0$. So again choose $x = 0$. In summary, the closest point on the curve is $(0, 0)$ in case $c \le 1/2$. If $c > 1/2$, there are two closest points, namely, $(\pm\sqrt{(2c - 1)/2}, (2c - 1)/2)$.

C26. The points A, B, C should be chosen so that $OA = OB = OC$. Let c be the constant sum of the lengths of the edges. Then by the AM-GM inequality we see that

$$\begin{aligned} c &= x + y + z + \sqrt{x^2 + y^2} + \sqrt{x^2 + z^2} + \sqrt{y^2 + z^2} \\ &\ge x + y + z + \sqrt{2xy} + \sqrt{2xz} + \sqrt{2yz} \\ &\ge 3\sqrt[3]{xyz} + 3\sqrt{2}\sqrt[3]{xyz} = (3 + 3\sqrt{2})\sqrt[3]{xyz} \end{aligned}$$

with equality holding throughout iff $x = y = z$.

C27. Choose H and K so that $HQ = KQ$ and $HK = c$. One way to see this is to start with a line segment HK of length c inscribed in a circle of the appropriate size so that $\angle HBK = \angle PQR$ for any point B on the arc HK. (Thus arc HK is a minor arc if $\angle PQR$ is obtuse, a major arc if $\angle PQR$ is acute, and a semicircular arc if $\angle PQR = 90°$.) The triangle HBK has largest area if B is the midpoint of the arc.

C28. K should be the midpoint of the line segment $A'B'$ where A' and B' are the feet of the perpendiculars from A and B to the line PQ. To prove this, write a, b, c for the lengths AA', BB', and $A'B'$. Let $A'K = x$ so that $KB' = c - x$; we prove that $x = c/2$ for the required minimum. Observe that

$$\begin{aligned} AK^2 + KB^2 &= a^2 + x^2 + b^2 + (c - x)^2 \\ &= a^2 + b^2 + c^2 + 2(x^2 - cx). \end{aligned}$$

C29. (i) Locate P at B. (ii) Locate P at any position on the line segment BC. (iii) If the number of points is odd, locate P at the middle one. If the number is even, locate P anywhere on the line segment joining the two middle ones. To prove this, think of P as a moving point, starting to the left of all the given points on the line. As P moves to the right the sum $PA + PB + PC$ in case (i) decreases until P reaches B, and then increases again. In case (ii) the sum $PA + PB + PC + PD$ decreases until P reaches B, then is a constant until P reaches C, then increases again as P continues to the right. This analysis is valid in case (iii) also.

C30. $\sqrt{620}$ feet. This can be obtained by two applications of the theorem of Pythagoras, first to get the diagonal distance $\sqrt{520}$ across the floor, because $520 = 14^2 + 18^2$. The answer then is the length of the hypotenuse in a right triangle with sides 10 and $\sqrt{520}$.

C31. 30 feet. Conceive of a side wall, of dimensions 10 by 18, as hinged along the 18 foot length at the floor, and hypothetically push it over so that it lies next to the floor. Thus we have a 10×18 rectangle attached to a 14×18 rectangle along the 18-foot side. Together these form a rectangle 18 by 24, with diagonal of length 30. This is the shortest path. To verify this, it is necessary to compare this result with two other possibilities for "unfolding" the room. One is to push over an end wall to create a rectangle 14 by 28. Another is to swivel an end wall around so it stands next to a side wall, giving a rectangle 10 by 32.

C32. If n is even, locate half the points at one end of the segment and the other half at the other end. If n is odd, say $n = 2k + 1$, locate k points at one end and k at the other end, and then place the final point arbitrarily on the segment. One way to prove this is to take the segment as the unit of length, so that the endpoints can be assigned the coordinates 0 and 1. Any n points have coordinates $a_1, a_2, \ldots, a_n$ with

$$0 \leq a_1 \leq a_2 \leq \cdots \leq a_n \leq 1$$

by the obvious ordering of the points. The problem is to minimize $\Sigma\,(a_j - a_i)$, where the sum is taken over all integer pairs, i,j satisfying $1 \leq i < j \leq n$. Removing parentheses and performing all possible cancellations we see that this sum reduces to

$$(n-1)a_n + (n-3)a_{n-1} + (n-5)a_{n-2} + \cdots + (3-n)a_2$$
$$+ (1-n)a_1 = \sum_{r=1}^{n} (2r - n - 1)a_r.$$

If n is even, there are as many terms a_r with positive coefficients as with negative coefficients, and the maximum is attained by taking $a_r = 0$ for $r = 1, 2, \ldots, n/2$, and $a_r = 1$ for all other values of r. If n is odd $n = 2k + 1$ say, then in the above sum the term a_r with $r = (n + 1)/2$ has coefficient 0 and so a_r can be assigned an arbitrary value from 0 to 1.

C33. No. Consider the case $a = 200$, $b = c = 101$, $a' = b' = c' = 100$. But area T' < area T holds if T is not an obtuse-angled triangle. On the other hand, if T has an obtuse angle, then there are triangles T' with shorter sides but larger area. To prove these assertions, first suppose that the angles α, β, γ of T are each 90° or less. If the corresponding angles of T are α', β', γ', then at least one of

$$\alpha' \leq \alpha, \quad \beta' \leq \beta, \quad \gamma' \leq \gamma$$

holds, since the sum of the angles of a triangle is constant. Without loss of generality we assume that $\alpha' \leq \alpha$. Then we see that

$$\text{area } T' = \frac{1}{2}b'c' \sin \alpha' < \frac{1}{2}bc \sin \alpha = \text{area } T.$$

Finally, suppose that T is an obtuse-angled triangle, say $\alpha > 90°$. Then $a^2 > b^2 + c^2$, and the right triangle with sides b, c, $(b^2 + c^2)^{1/2}$ has area larger than T by Theorem 3.2c. If T' is the triangle with sides

$$rb, \quad rc, \quad r(b^2 + c^2)^{1/2}$$

with $r < 1$ but r very close to 1, then area T' > area T.

C34. 4 by 8; $128\sqrt{12}/9$. The vertices of the rectangle can be taken as $(\pm x, 0)$ and $(\pm x, 12 - x^2)$, and hence the problem is to maximize the area $2x(12 - x^2)$. Writing the square in the form $2(2x^2)(12 - x^2)(12 - x^2)$, we use Theorem 2.6a to write $2x^2 = 12 - x^2$, giving $x = 2$. As to the trapezoid, if the upper vertices are $(\pm x, 12 - x^2)$ then the area to be maximized is

$$(12 - x^2)(\sqrt{12} + x), \quad \text{or} \quad (2\sqrt{12} - 2x)(\sqrt{12} + x)(\sqrt{12} + x)/2.$$

Again we use Theorem 2.6a, with the result

$$2\sqrt{12} - 2x = \sqrt{12} + x.$$

C35. Ratio $\sqrt{2}$. To prove this, we note that we can just as well minimize the square

$$S^2 = \pi^2 r^2(r^2 + h^2) = \pi^2(r^4 + 9V^2/\pi^2 r^2)$$
$$= \pi^2(r^4 + 9V^2/2\pi^2 r^2 + 9V^2/2\pi^2 r^2).$$

The terms of this sum have a constant product; so S is minimized by taking

$$r^4 = 9V^2/2\pi^2 r^2, \quad 2\pi^2 r^6 = 9V^2 = \pi^2 r^4 h^2, \quad h = r\sqrt{2}.$$

C37. Without loss of generality we may assume that $a < b$. It is not difficult to show that the only trapezoids that need be considered are those such that the perpendiculars from the vertices of the side of length a to the side of length b lie inside the trapezoid. Suppose that these perpendiculars separate the base b into three segments of lengths x, a, and $b - a - x$. Then the perimeter is seen to be

$$a + b + \sqrt{x^2 + h^2} + \sqrt{(b - a - x)^2 + h^2}.$$

By the preceding problem with $a_3 = b_3 = 0$ and a_1, b_1, a_2, b_2 replaced by x, h, $b - a - x$, and h, we see that the perimeter is at least

$$a + b + \sqrt{(b - a)^2 + (2h)^2}$$

with equality if x and h are proportional to $b = a - x$ and h. This leads to $x = b - a - x$ or $x = (b - a)/2$, giving the isosceles trapezoid.

C38. $x = (bc - ad)/(a + c)$. By Problem C36 with $a_3 = b_3 = 0$, the sum is $\geq \sqrt{(a + c)^2 + (b + d)^2}$ with equality iff the pairs a, $b - x$ and c, $d + x$ are proportional, giving the solution as shown. (Depending on the values a, b, c, d the fraction $(bc - ad)/(a + c)$ may be positive, negative, or zero, but $b - x$ and $d + x$ are positive nevertheless, as they must be to satisfy the proportionality condition.)

C39. Any point Q in the interior or on the boundary of the polygon will do. For if the regular polygon has n sides of length a, and if the distances to the sides from any such point Q are $d_1, d_2, \ldots, d_n$, then by joining Q to the vertices we see that the area of the polygon is

$$ad_1/2 + ad_2/2 + \cdots + ad_n/2.$$

Hence the sum $d_1 + d_2 + \cdots + d_n$ is a constant for every such point Q.

C40. Consider a circle tangent to OQ and OR at H and K, respectively, as in Figure C40, such that H, K, and P are collinear. For any other line segment H_1PK_1 through P we see that $PH \cdot PK < PH_1 \cdot PK_1$ by the standard theorem about two intersecting chords of a circle. (The theorem states that if P is the intersection point of two chords HK and AB, then $PH \cdot PK = PA \cdot PB$.)

C41. 800 meters. When A has moved a distance x, cyclist B has moved $4x/3$, with a distance d between them given by

$$d^2 = (1000 - x)^2 + 16x^2/9,$$

as noted from Figure C41. This gives $d^2 = 10^6 - 2000x + 25x^2/9$. The smallest value of d^2 occurs when $x = 360$ by Corollary 1 of Theorem 2.2a.

C42. The area of any quadrilateral $ABCD$, with vertices in that order, is at most $\frac{1}{2}AC \cdot BD$. To see this, observe that the sum of the altitudes from B and D to the base AC in the triangles ABC and ADC is at most BD. (This argument is valid even for a nonconvex quadrilateral if the vertices are labeled so that C lies inside the triangle ABD.) If the diameter of the quadrilateral is 1, then $AC \leq 1$ and $BD \leq 1$. To exhibit quadrilaterals with diameter 1 and area ½ other than the square, consider any two perpendicular, intersecting line segments AC and BD, each of length 1, such that the length of each side of $ABCD$ is at most 1.

D1. (a) Yes. By joining the vertices to the center of the circle, we find that the triangles so formed are congruent, and hence the interior angles of the polygon are equal. (b) Yes, if n is odd. If n is even, the polygon need not be regular, as in the case of a rhombus. Joining the center of the circle to the vertices of the n-gon and to the points of tangency, we have $2n$ triangles, of which n are congruent, and the other n are also congruent. This implies that the interior angles of the polygon have an $a, b, a, b, a, b, \ldots$ pattern. Thus the angles are equal if n is odd.

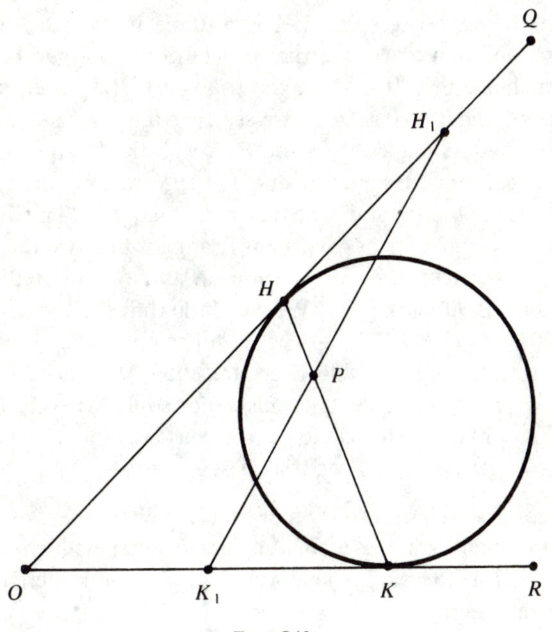

FIG. C40

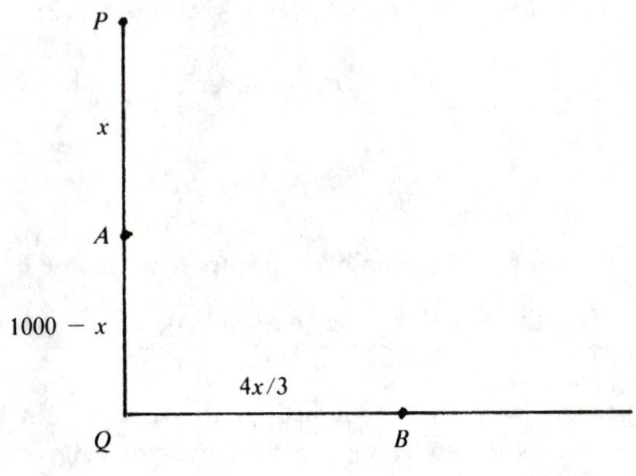

FIG. C41

D2. (a) Yes, if n is odd. If the number n of sides is even, the polygon need not be regular, as for example in the case of a rectangle inscribed in a circle. Let the vertices of the polygon be $P_1, P_2, \ldots, P_n$ in that order. If $n = 3$, the triangle $P_1P_2P_3$ is equiangular and hence equilateral. If $n > 3$, we see that the segments P_1P_3 and P_2P_4 subtend equal angles at P_2 and P_3, respectively, and so $P_1P_3 = P_2P_4$. Thus the triangles $P_1P_2P_3$ and $P_4P_3P_2$ are congruent, having two equal pairs of sides and an equal pair of angles. (This is the so-called "ambiguous" case of congruent triangles, because the angle is not included between the sides. However, there is no ambiguity in case the angle is 90° or larger.) We conclude that $P_1P_2 = P_3P_4$, and similarly $P_2P_3 = P_4P_5$, $P_3P_4 = P_5P_6$, and so on. Thus, of any three consecutive sides, the first and third are equal. If n is odd this implies that all sides are equal, but this is not so if n is even. (b) Yes. Joining the center of the circle to the vertices and the points of tangency, we get $2n$ congruent triangles.

D3. If some arc from A to B other than the circular arc of length c encloses an equal or larger area, this arc together with the "rest of the circle," as in the suggestion, would give a contradiction to the isoperimetric theorem.

D4. $\theta = 2$. To prove this we note that the isoperimetric quotient is

$$4\pi A/L^2 = 2\pi r^2\theta/(2r + r\theta)^2 = 2\pi\theta/(\theta + 2)^2.$$

To maximize this, we minimize the inverse with the constant 2π omitted,

$$(\theta + 2)^2/\theta = \theta + 4/\theta + 4.$$

By Example 2 in §2.2 the minimum occurs here in case $\theta = 2$.

D6. (i) Yes; the length of the diagonal AC of the parallelogram must be $(\sqrt{2} - 1)(AB + BC)$. (ii) No.

E1. The first inequality is implied by the result (1) in §5.4. The second can be obtained from the first by replacing α and β by $90° - \alpha$ and $90° - \beta$.

E2. Among acute-angled triangles the maximum is $3/2$ by Theorem 5.4a. For other triangles we prove that the sum is less than $3/2$. Any triangle has two acute angles, say α and β. It is clear that

$$\begin{aligned}
&\cos\alpha + \cos\beta + \cos\gamma \\
&\qquad = \cos\alpha + \cos\beta - \cos(\alpha + \beta) \\
&\qquad = \cos\alpha + \cos\beta - \cos\alpha\cos\beta + \sin\alpha\sin\beta \\
&\qquad = 1 + \sin\alpha\sin\beta - (1 - \cos\alpha)(1 - \cos\beta) \\
&\qquad < 1 + \sin\alpha\sin\beta.
\end{aligned}$$

By the preceding problem, $\sin\alpha\sin\beta \le 1/2$ for acute angles α and β with sum $\le 90°$.

To prove that the greatest lower bound is 1, note first that we can get as close to 1 as we please by taking $\gamma = 90°$, α small, and $\beta = 90° - \alpha$. Now for any acute angle α, $\sin\alpha > \sin^2\alpha$ and $\cos\alpha > \cos^2\alpha$, so $\sin\alpha + \cos\alpha > 1$. Hence $\sin\alpha > 1 - \cos\alpha$ and $\sin\beta > 1 - \cos\beta$ and $\sin\alpha\sin\beta > (1 - \cos\alpha)(1 - \cos\beta)$. This with the analysis above shows that $\cos\alpha + \cos\beta + \cos\gamma > 1$ for all triangles.

E4. $3\sqrt{3}/8$; $1/8$. The first follows from Problem E1 and Theorem 5.3b with $n = 3$ and the inequalities reversed. To prove the result $1/8$ for the product of the cosines, observe first that this product is zero or negative if one of the angles is $90°$ or more. Hence we confine attention to acute-angled triangles, and again we use Problem E1 and Theorem 5.3b with $n = 3$.

E5. $2\sqrt{3}$. This follows from inequality (2) in §5.4 and Theorem 5.3a with $n = 3$ and the inequalities reversed.

E6. If $\alpha + \beta = 90°$, then $\tan\beta = \cot\alpha = (\tan\alpha)^{-1}$, and hence there is equality in $(*)$. Otherwise we use the identity

$$\tan(\alpha + \beta) = (\tan\alpha + \tan\beta)/(1 - \tan\alpha\tan\beta)$$

and its variation

$$\tan(\alpha + \beta) = [2\tan(\alpha + \beta)/2]/[1 - \tan^2(\alpha + \beta)/2].$$

Since $\tan\alpha + \tan\beta \geq 2\tan(\alpha+\beta)/2 > 0$, it follows that

$$1 - \tan\alpha\tan\beta \geq 1 - \tan^2(\alpha+\beta)/2$$

provided $\tan(\alpha+\beta)$ is positive, that is, provided $\alpha+\beta < 90°$. If $\alpha + \beta > 90°$, then $\tan(\alpha+\beta) < 0$ and the inequality is reversed.

E7. Using the preceding problem we get

$$\tan(\alpha/2)\tan(\beta/2) \leq \tan^2(\alpha+\beta)/4$$

and

$$\tan(\gamma/2)\tan 30° \leq \tan^2(\gamma + 60°)/4.$$

If we multiply these, and observe that

$$\begin{aligned}\tan(\alpha+\beta)/4\tan(\gamma+60°)/4 &\leq \tan^2(\alpha+\beta+\gamma+60°)/8\\ &= \tan^2 30° = 1/3,\end{aligned}$$

the first result becomes clear. Turning to the second part, we may assume that the acute angles satisfy

$$\alpha \leq \beta \leq \gamma < 90°, \text{ so that } \gamma \geq 60° \text{ and } \alpha+\beta > 90°.$$

Then the preceding problem implies that

$$\begin{aligned}&\tan\alpha\tan\beta \geq \tan^2(\alpha+\beta)/2,\\ &\tan\gamma\tan 60° \geq \tan^2(\gamma+60°)/2,\\ &\tan(\alpha+\beta)/2\tan(\gamma+60°)/2 \geq \tan^2(\alpha+\beta+\gamma+60°)/4\\ &\qquad = \tan^2 60° = 3.\end{aligned}$$

The result for the sum $\tan\alpha + \tan\beta + \tan\gamma$ follows from Theorem 5.4a.

E8. $3\sqrt{3}/8$. We can restrict attention to the case $x + y < \pi$, since otherwise $\sin(x+y)$ is not positive. From Problem E1 we have $\sin x \sin y \leq 2\sin^2(x+y)/2$; so the maximum occurs if $x = y$. Hence the problem is to maximize $\sin^2 x \sin 2x$ with $0 < x < \pi/2$. Using $\sin 2x = 2\sin x\cos x$, and squaring for convenience, we want to maximize

$$4\sin^6 x\cos^2 x = \frac{4}{3}(\sin^2 x)(\sin^2 x)(\sin^2 x)(3 - 3\sin^2 x).$$

If we ignore the $4/3$, the terms of this product have a constant sum, so the maximum occurs in case $\sin^2 x = 3 - 3\sin^2 x$.

E9. The answer is the same in both cases: Locate R so that $PR = QR$, with R on the major arc PQ. To maximize the area, take PQ as the base of the triangle and maximize the altitude. As to the perimeter, for any location R_1 on the major arc PQ the perimeter of PQR_1 is $2r(\sin\alpha + \sin\beta + \sin\gamma)$, where the angles subtended at the center by the arcs PQ, QR, and RP, respectively, are taken as 2α, 2β, and 2γ. Now α is a constant, and $\beta = \gamma$ follows from Theorem 5.2a.

E10. The first is true but the converse is not. Using the law of sines in the form

$$a = k\sin\alpha, \quad b = k\sin\beta, \quad c = k\sin\gamma$$

for some k, we see that $a < (b + c)/2$ implies

$$\sin\alpha \le (\sin\beta + \sin\gamma)/2.$$

This with Theorem 5.2a gives $\sin\alpha < \sin(\beta + \gamma)/2$, and this implies

$$\alpha < (\beta + \gamma)/2.$$

To prove the converse false, we note that the triangle with sides $\sqrt{3}$, 1, 2 has angles 60°, 30°, 90°, and that $\sqrt{3} > (1 + 2)/2$. So the triangle with α just a little smaller than 60°, β just larger than 30°, and $\gamma = 90°$ satisfies $\alpha < (\beta + \gamma)/2$, whereas $a < (b + c)/2$ is false.

E11. $\sqrt{2}$.

E12. 5. The problem amounts to finding the maximum of $4\cos\theta + 3\sin\theta$ by writing $x = \cos\theta$.

E13. $\sqrt{51}$. The expression can be written as

$$4\sqrt{3}(x/\sqrt{3}) + \sqrt{3}\cdot\sqrt{1 - x^2/3} = 4\sqrt{3}\cos\theta + \sqrt{3}\sin\theta$$

by using $\cos\theta = x/\sqrt{3}$.

E14. 9. We can use the second part of Theorem 5.5a by putting the expression into the form (3); thus

$$5\sqrt{9 + 4x^2} - 8x = 8[\tfrac{5}{4}\sqrt{9/4 + x^2} - x].$$

Hence we apply the theorem with $a = \frac{5}{4}$ and $c = \frac{3}{2}$ to get $x = c/\sqrt{a^2 - 1} = 2$.

E15. The answer is max(a, b), as can be seen by looking at the expression in the form $a + (b - a)\sin^2\theta$ and making use of the fact that the maximum and minimum values of $\sin^2\theta$ are 1 and 0.

E16. (a) $4\sqrt{3}/9$; (b) $17/18\sqrt{6}$; (c) $10/3$. (a) Write $\sin\theta \sin 2\theta$ as $2\cos\theta\,(1 - \cos^2\theta)$. The square of this can be written as

$$2(2\cos^2\theta)(1 - \cos^2\theta)(1 - \cos^2\theta),$$

where the three factors in parentheses have a constant sum; so we take $2\cos^2\theta = 1 - \cos^2\theta$. (b) Write $\sin\theta\cos 2\theta$ as

$$\sin\theta\,(1 - 2\sin^2\theta),$$

and proceed as in part (a). (c) Write the expression as

$$3 + 2\cos\theta - 3\cos^2\theta.$$

This amounts to maximizing $3 + 2x - 3x^2$; so we set $x = 1/3$ as in §2.2.

E17. 4. Since $\tan\theta\cot\theta = 1$, this amounts to minimizing $x + 4/x$.

E18. 6; 10. Since $\sin\theta\operatorname{cosec}\theta = 1$, the first part amounts to minimizing $9x + 1/x$, leading to $x = 1/3$. In the second part we seek the minimum of $x + 9/x$ among positive values of x with $x \leq 1$. So we take $x = 1$ by Problem B7.

F1. In Figure 6.5a the area of triangle CPQ is

$$\frac{1}{2}CP\cdot CQ\cdot\sin\theta = \frac{1}{2}\sin\theta.$$

Since the area of the sector CPQ is $\theta/2$, the first result follows. To prove (ii) we first note that the inequality $\cos\theta > 1 - \theta$ is obvious if $1 < \theta < \pi/2$ because $\cos\theta$ is positive and $1 - \theta$ is negative for any such angle. If $0 < \theta \leq 1$, we use $\sin^2\theta + \cos^2\theta = 1$. Suppose $\cos\theta \leq 1 - \theta$. Then we see that

$$1 = \sin^2\theta + \cos^2\theta < \theta^2 + (1 - \theta)^2 = 1 - 2\theta + 2\theta^2.$$

This gives $0 < -2\theta + 2\theta^2$, which implies $\theta < \theta^2$, and this is false for $0 < \theta \leq 1$.

F2. The points should be spaced uniformly, so that if points P and S are at the ends of the diameter, then Q and R should be located so that PQ, QR, and RS all subtend angles of 60° at the center. To prove this, suppose that in general the sides of the quadrilateral, other than PS, subtend angles α, β, γ at the center, with $\alpha + \beta + \gamma = 180°$. Then the area of the quadrilateral is $\frac{1}{2}r^2(\sin\alpha + \sin\beta + \sin\gamma)$, and this is a maximum if $\alpha = \beta = \gamma$.

F3. $\sqrt{2}$, $1/2$, and 5. First, since $x^2 + y^2$ is a constant, the maximum value of x^2y^2 occurs if $x^2 = y^2$, and likewise for the maximum of xy. Then, since the relation $(x + y)^2 = x^2 + 2xy + y^2 = 1 + 2xy$ holds for all x and y under consideration, the maximum value of $x + y$ also occurs if $x = y$.

A more general way of dealing with $x + y$ is to observe that $x + y = c$ represents a family of parallel lines, one line for each value of the constant c. To maximize $x + y$, we ask the question: what is the largest value of c so that the line $x + y = c$ has a point in common with the circle $x^2 + y^2 = 1$? The answer of course is to choose c so that $x + y = c$ is a tangent line to the circle, and we get the line $x + y = \sqrt{2}$ by simple geometry.

In a similar way we find the maximum value of $3x + 4y$ on the circle by looking for the positive value of c so that $3x + 4y = c$ is a tangent line to the circle. Such a line has slope $-3/4$; so the perpendicular line through the origin has slope $4/3$, and equation $4x - 3y = 0$. This line intersects the circle $x^2 + y^2 = 1$ at the point $(3/5, 4/5)$; so this is the point of tangency. To find c we substitute these coordinates into $3x + 4y = c$ to get $c = 9/5 + 16/5 = 5$. Thus the maximum value of $3x + 4y$ on $x^2 + y^2 = 1$ is 5.

F4. If n is even, say $n = 2k$, we can take each of the points $(1, 0)$ and $(-1, 0)$ exactly k times. If n is odd, say $n = 2k + 1$, we can take $(1, 0)$ exactly k times, $(-1, 0)$ exactly $k - 1$ times, and each of $(-1/2, \sqrt{3}/2)$ and $(-1/2, -\sqrt{3}/2)$ once. There are other ways of selecting the points, for example by taking the n points uniformly spaced around the circle. This works for both n even and n odd.

These solutions are suggested by the following theory, which serves also to show that the sum n^2 cannot be exceeded. Consider any n points (x_i, y_i), $i = 1, 2, \ldots, n$, on the circle $x^2 + y^2 = 1$. The sum S of the squares of all distances between pairs of these points is

$$S = \Sigma\,[(x_i - x_j)^2 + (y_i - y_j)^2],$$

where the sum is over all pairs of integers i,j satisfying $1 \leq i < j \leq n$. Expanding these squares we note that if $(\Sigma\, x_i)^2 + (\Sigma\, y_i)^2$ is added to both sides of this equation, there is a cancellation of all cross-product terms on the right to give

$$S + (\Sigma\, x_i)^2 + (\Sigma\, y_i)^2 = n\,\Sigma\, x_i^2 + n\,\Sigma\, y_i^2, \tag{1}$$

where each sum here is taken over $i = 1, \ldots, n$. Now $x_i^2 + y_i^2 = 1$ for all values of i, so the right side of equation (1) is just n^2. Hence we have

$$S = n^2 - (\Sigma\, x_i)^2 - (\Sigma\, y_i)^2.$$

This shows that $S > n^2$ is impossible, and also that $S = n^2$ is achieved by choosing the points so that $\Sigma\, x_i = \Sigma\, y_i = 0$. This leads to the kinds of solutions given in the preceding paragraph.

G1. Parallel lines are carried into parallel lines, but perpendicular lines are not carried into perpendicular lines in general. But the perpendicular lines $x = c_1$ and $y = c_2$ are carried into the perpendicular lines $X = c_1/\mathrm{a}$ and $Y = c_2/\mathrm{b}$. The explanation is simple. Any line $y = mx + k$ with slope m is mapped onto the line $bY = maX + k$ with slope ma/b. Hence two parallel lines with slope m are mapped onto two lines with slopes ma/b, hence parallel. But consider two perpendicular lines with slopes m_1 and m_2. The condition for perpendicularity is $m_1 m_2 = -1$. The two lines are mapped onto lines with slopes $m_1 a/b$ and $m_2 a/b$. The product of these two slopes is $-a^2/b^2$, and this is not -1.

G2. Let (x, y) be any point on the ellipse $x^2/a^2 + y^2/b^2 = 1$ with $x \neq 0$ and $y \neq 0$. Then the four points $(\pm x, \pm y)$ lie on the ellipse and form a rectangle with sides parallel to the x axis and the y axis, and so the sides are parallel to the axes of the ellipse.

On the other hand, suppose that the points A, B, C, D on the ellipse form a rectangle $ABCD$, and that the sides of the rectangle are not parallel to the x and y axes. We want to prove that this is impossible. Apply the mapping $x = aX$, $y = bY$, so that the ellipse $x^2/a^2 + y^2/b^2 = 1$ is transformed into the circle $X^2 + Y^2 = 1$. Let the points A, B, C, D be mapped onto the points A', B', C', D' on the circle. By the preceding problem, the parallel sides of $ABCD$ are mapped onto parallel sides of $A'B'C'D'$, but the perpendicularity is lost, so that A', B', C', D' is a parallelogram but not a rectangle. But it is impossible to inscribe a parallelogram in a circle, because the sum of a pair of opposite angles of a quadrilateral inscribed in a circle is 180°.

G3. In case $\theta = \pi/4$ the four points are given by formulas (2). The quadrilateral formed by these four points has two sides parallel to the x axis and two sides parallel to the y axis. Hence it is a rectangle. For other values of θ we argue as follows. Write P, Q, R, S for $(a \cos \theta, b \sin \theta)$, $(-a \sin \theta, b \cos \theta)$, $(-a \cos \theta, -b \sin \theta)$, $(a \sin \theta, -b \cos \theta)$, respectively. It is not difficult by the use of slopes of lines to prove that PQ and RS are parallel, as also are PS and QR. But PQ is not perpendicular to PS because the product of their slopes turns out to be $-b^2/a^2$, which is not -1 because we are assuming throughout that $a > b$.

G4. For the triangle, $3\sqrt{3}ab/4$; for the n-gon, $\frac{1}{2}nab \sin 2\pi/n$. These results follow from the facts that the area of an equilateral triangle inscribed in the unit circle is $3\sqrt{3}/4$ and that the area of a regular n-gon inscribed in the unit circle is $\frac{1}{2}n \sin 2\pi/n$. Formula (3) of §7.1 is also helpful.

G5. 2. A line with slope m through the point $(4, -3)$ has equation $y + 3 = m(x - 4)$ or $y = mx - 4m - 3$. Eliminating y between this equation and $x^2 - 4y - 28 = 0$ gives $x^2 - 4mx + 16m - 16 = 0$. The roots of this equation are equal iff $(4m)^2 - 4(16m - 16) = 0$, with solution $m = 2$.

G6. $(0, 1)$ is closest and $(0, -1)$ is farthest. The square of the distance from $(0, 3)$ to (x, y) on the curve is

$$\begin{aligned}(x - 0)^2 + (y - 3)^2 &= (2 - 2y^2) + y^2 - 6y + 9 \\ &= 11 - (y^2 + 6y).\end{aligned}$$

We want the largest and smallest values of this among values of y from $y = -1$ to $y = +1$, because these are the smallest and largest values of y on the curve. Now $y^2 + 6y = (y + 3)^2 - 9$, and this increases as y goes from $y = -1$ to $y = +1$.

G7. If $k \geq 16/5$, the answer is (5,0). If $k < 16/5$, there are two closest points with coordinates $x = 25k/16$ and $y = \pm 3\sqrt{1 - 25k^2/256}$. To prove this, we look at the square of the distance from (5,0) to any point (x, y) on the curve, namely,

$$(x - k)^2 + y^2 = k^2 + 9 - \frac{16}{25}(25kx/8 - x^2).$$

To maximize this we want to maximize $25kx/8 - x^2$, which leads to $x = 25k/16$ by Theorem 2.2a. But x is a coordinate of a point on the curve $x^2/25 + y^2/9 = 1$; so the largest possible value of x is $x = 5$. So (5,0) is the answer if $25k/16 \geq 5$, that is, if $k \geq 16/5$, where we are using the simple idea of problem B7 about increasing and decreasing quadratic polynomials. On the other hand, if $k < 16/5$, the value $x = 25k/16$ gives the location of the closest points.

G8. The points $(3, \pm 3\sqrt{2})$. If we take the square of the distance from the origin to a point (x, y) on the curve, we have

$$x^2 + y^2 = x^2 + 54/x = x^2 + 27/x + 27/x.$$

This sum has a constant product, so the minimum occurs if

$$x^2 = 27/x \text{ or } x^3 = 27.$$

G9. Nearest to (3,0) is (1,0); nearest to (6,0) are $(2, \pm\sqrt{6})$. In the first case the problem is to minimize

$$(x - 3)^2 + y^2 \text{ or } 3x^2 - 6x + 7$$

by use of $y^2 = 2x^2 - 2$. This leads to $x = 1$. In the second case the expression to be minimized is

$$(x - 6)^2 + y^2 \text{ or } 3x^2 - 12x + 34,$$

and so we get $x = 2$.

G10. Here is a proof that the two points are $(a, 0)$ and $(-a, 0)$. First if P and Q are two points on the ellipse so that the origin (0, 0) is not on the line segment PQ, then one at least of PP' and QQ' is longer than PQ, where P' and Q' are the points on the ellipse diametrically opposite P and Q. (That is, the lines PP' and QQ' pass through the origin.) Hence the problem amounts to finding the points P on the ellipse to maximize the distance PO, where O is $(0,0)$.

If P has coordinates (x, y), then

$$PO^2 = x^2 + y^2 = a^2 - (a^2 - b^2)y^2/b^2.$$

Since $a^2 - b^2 > 0$, the maximum value of this is clearly a^2, by setting $y = 0$.

G11. $2\sqrt{a^4 + b^4}$. As argued in G10 above, the problem amounts to finding a point P on the curve farthest from the origin. Thus we want to maximize $x^2 + y^2$ subject to the condition $x^4/a^4 + y^4/b^4 = 1$. Writing u for x^2 and v for y^2, the question is to maximize $u + v$ subject to $u^2/a^4 + v^2/b^4 = 1$. The answer to this question is $\sqrt{a^4 + b^4}$ by Example 1 in §7.5.

G12. Solving the equation $2x^2 + 2xy + y^2 + 4x + 4y - 14 = 0$ for y in terms of x, we find that

$$y = -x - 2 \pm \sqrt{-x^2 + 18}.$$

The largest and smallest values of x that can be used here are $x = \sqrt{18}$ and $x = -\sqrt{18}$. Hence the answers are $(\sqrt{18}, -\sqrt{18} - 2)$ and $(-\sqrt{18}, \sqrt{18} - 2)$.

G13. Since the equation of the ellipse is the special case with $k = 9$ in equation (5) in the text, we can use equations (6) in this case. Thus we have

$$x = y + 5 \pm \sqrt{-(y - 3)^2 + 25}.$$

The expression under the square-root sign equals zero for two values of y, namely, $y = 8$ and $y = -2$. These are the largest and smallest values of y among all points on the ellipse, corresponding to the points $(8,13)$ and $(-2, 3)$. The center of the ellipse is the midpoint of the line segment joining these two symmetrically located points. Hence the center is $(3,8)$.

H1. Place a congruent replica adjacent to the triangle itself to form a parallelogram, which then obviously serves to tile the plane.

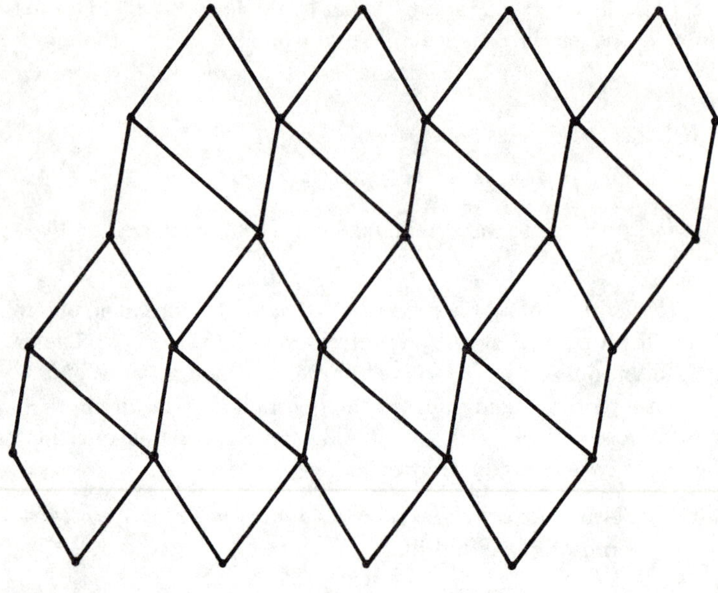

Fig. H2a

H2. In the case of a convex quadrilateral place two copies of the quadrilateral in adjacent positions, joined at a side of the same length, as in Figure H2a. Thus, ignoring the common side, we have formed a hexagon whose opposite sides are parallel and equal. This hexagon can then be used to tile the plane. In the case of a non-convex quadrilateral, the same procedure works, as shown in Figure H2b. The difference in this case is that the hexagon formed is non-convex, but again it is easy to see that it tiles the plane.

H3. As in Figure H3 remove the triangle QRS from the pentagon to create a hexagon. Let A and L denote the area and perimeter of the pentagon, and let the angle at Q be 2θ, so that $\theta < 90°$. If $QR = QS = x$, then the area and the perimeter of the hexagon are

$$A - \frac{1}{2}x^2 \sin 2\theta \quad \text{and} \quad L - 2x + 2x \sin \theta,$$

respectively. Thus we want to prove that

$$A/L^2 < \left(A - \frac{1}{2}x^2 \sin 2\theta\right)/(L - 2x + 2x \sin \theta)^2.$$

Cross-multiplying and canceling the term AL^2, this amounts to

$$\frac{1}{2}L^2 x \sin 2\theta + 4Ax(1 - \sin \theta)^2 < 4AL(1 - \sin \theta),$$

after a little algebraic simplification. This inequality certainly holds if x is small enough.

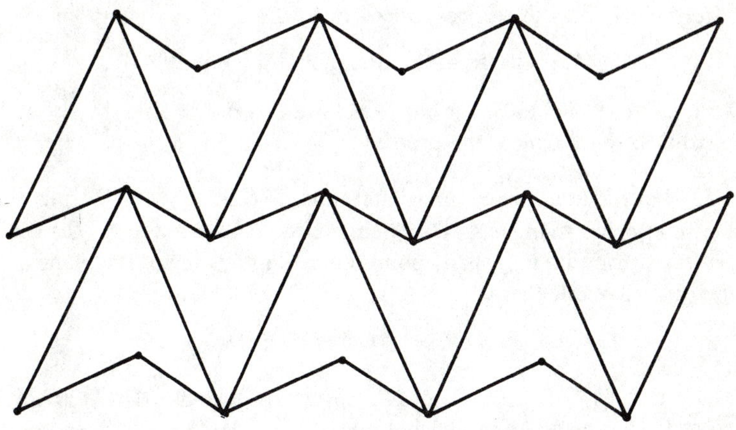

Fig. H2b

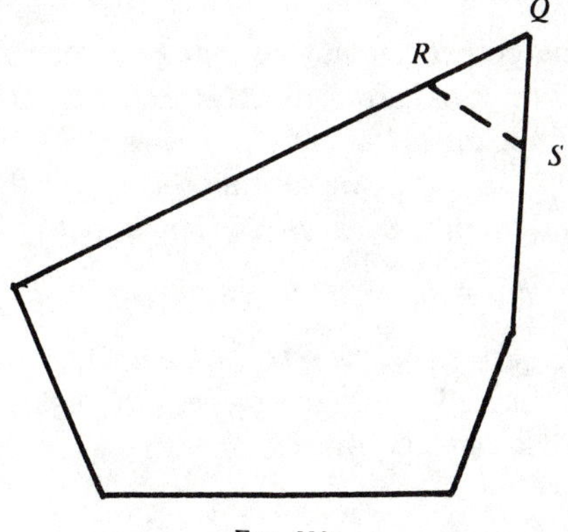

Fig. H3

I1. Locate P at A. To prove this we must show that, if P is located at any position other than A, then $PB + PC < AB + AC$. This is obvious if P lies on BC. If P lies on AB, then $AB + AC = PB + PA + AC > PB + PC$. If P lies on AC, a similar argument applies. If P lies in the interior of the triangle, join CP and extend it to intersect AB at Q. Then $AB + AC > QB + QC$ and also

$$QB + QC = QB + QP + PC > PB + PC.$$

I2. Locate P at the intersection of the diagonals. For if Q is any other point, and the vertices are A, B, C, D in that order, then

$$QA + QC \geq AC \quad \text{and} \quad QB + QD \geq BD,$$

with a strict inequality in at least one case. Adding these inequalities, we complete the proof.

I3. Let the nonconvex quadrilateral be $ABCD$, say with D lying in the interior of triangle ABC. Then P should be located at D. We want to prove that if Q is any point other than D lying in the plane of the quadrilateral, then

$$QA + QB + QC + QD > DA + DB + DC.$$

The triangles QAB, QBC, QAC, cover the triangle ABC; so the point D lies inside or on the boundary of at least one of these, say QBC. Then $QB + QC > DB + DC$ by problem **I1**. Add this inequality to $QA + QD \geq DA$.

I4. If CP is the perpendicular line segment from C to AB, then $CK \geq CP$, so it suffices to prove that $AB + CP > AC + BC$. First consider the case where $\angle C = 90°$. Denoting the lengths AB, AC, BC, and CP by c, b, a, and p, we see that the area of the triangle can be expressed as $cp/2$ and also as $ab/2$. Hence $cp = ab$ and

$$c^2 + pc = c^2 + ab = (c - a)(c - b) + ac + bc > ac + bc.$$

Dividing by c we get $c + p > a + b$. Second, consider the case where $\angle C > 90°$. Let Q be chosen on the line segment AB so that $\angle QCB = 90°$. Then we can write

$$AB + CP = AQ + (QB + CP) > AQ + (CQ + CB) > AC + BC.$$

I5. One at least of the angles CPA and CPB is $90°$ or more, and hence at least one of $CP < CA$ or $CP < CB$ holds.

I6. The point P should be located at a vertex with a smallest angle. If there is a unique smallest angle, say $\angle A$, then there is a unique location for P, namely at A. Otherwise P is located at either A or B in case $\angle A = \angle B < \angle C$, and P is located at any vertex in the case of an equilateral triangle. A proof is given here in the case of a unique smallest angle at A; the other cases follow without difficulty. We want to prove that if P is located inside or on the triangle ABC, but not at A, then $AB + AC > PA + PB + PC$. If P is at B or at C, this inequality follows readily. If P lies in the interior of one of the three sides, the inequality follows by an application of the preceding problem.

Finally, suppose that P lies in the interior of the triangle. Then at most one of the angles APB, APC, BPC, is acute, for if two of these were acute angles the third would exceed $180°$. So we can be certain that at least one of the angles APB and APC is $90°$ or more. We treat the case where $\angle APB \geq 90°$; the other case is similar. Join CP and extend it to intersect AB at K. Then by Problem I4 we have $PK + AB > PA + PB$. By I5 we conclude that

$$CK < \max(AC, BC) = AC.$$

It follows that

$$AC + AB > CK + AB = PC + PK + AB > PC + PA + PB.$$

I7. The equilateral triangle is special, because in this case the point P can be located anywhere inside or on the boundary of the triangle. In other cases the point P is located at a vertex of the triangle having a largest angle. If there is a unique largest angle, there is a unique location for P; if two largest angles, then two locations for P.

To prove these results, let p_1, p_2, p_3 be the lengths of the perpendiculars from any point P to the sides AB, AC, BC, respectively. (We presume that P lies inside or on the boundary of the triangle, because no location outside the triangle can provide a solution of the problem.) Let the lengths of the sides AB, AC, BC be denoted by the usual c, b, a, and let h be the altitude from a largest angle, say angle A. Then the area of the triangle can be expressed in two ways,

$$\tfrac{1}{2}ha = \tfrac{1}{2}(p_1c + p_2b + p_3a).$$

In the case of an equilateral triangle, $a = b = c$ and so $p_1 + p_2 + p_3 = h$. Otherwise we have

$$a = \max(a, b, c) > \min(a, b, c)$$

and it follows that $h < p_1 + p_2 + p_3$.

I8. The points P and Q should be located to satisfy the following conditions: $AP = DP$, $BQ = CQ$, the three angles at P should be 120°, and likewise the three angles at Q. To prove this, take P and Q in arbitrary locations as in Figure I8. Consider the line through P parallel to AD: by Heron's theorem of §3.5 the point P' on this line that is equidistant from A and D has the minimum sum $AP' + P'D$. Likewise, among points on the line through Q parallel to BC the point Q' has the minimum sum $BQ' + Q'C$. Also $P'Q' \leq PQ$, and so we have

$$AP' + DP' + P'Q' + BQ' + CQ' \leq AP + DP + PQ + BQ + CQ,$$

with equality iff P coincides with P', and Q with Q'. The rest of the conclusion, namely, the 120° angles part, follows from Theorem 9.2a.

I9. Each of AA', BB', CC' equals the sum $AP + BP + CP$, where P is the Fermat point of the triangle.

I10. Yes, it is always true. Add the inequalities

$$AR + AQ > QR, \quad BR + BP > PR, \quad CP + CQ > PQ.$$

I11. No, it is not always true. It *is* true if ABC is an equilateral triangle; this follows from Problem I5. If ΔABC is not equilateral, the result holds for some locations of K and not for others. An example of a location for the point K so that the result fails can be obtained as follows for any nonequilateral triangle. Label the vertices of such a triangle so that A is a smallest angle and AB is a longest side. Then locate K near both the vertex A and the side AB, so that AP and BQ are very close to AB in length, and CR close to CA.

I12. Take the coordinates of the vertices of any triangle as (0, 0), $(6a, 0)$ and $(6b, 6c)$. The centroid C is $(2a + 2b, 2c)$ and the area is $18ac$. Take the point Q with coordinates $(t, 0)$ on the line segment from $(3a, 0)$ to $(6a, 0)$. The line QC intersects the triangle again at R, say, with coordinates $(2bt/(t - 2a), 2ct/(t - 2a))$. The area of the triangle OQR, where O is the origin, is $ct^2/(t - 2a)$. Dividing the total area $18ac$ by this we get $18a(t - 2a)/t^2$ or $18a(t^{-1} - 2at^{-2})$. The largest and smallest values of this are 9/4 and 2 for values of t from $3a$ to $6a$, by use of Theorem 2.2a and Problem B7.

I13. $s = \frac{1}{2}$ and $s = 2$. From the preceding solution we get $RC/CQ = 2a/(t - 2a)$ by comparing the y coordinates of the points. Use $3a \le t \le 6a$.

I14. Following the suggestion, consider the triangle with vertices $(ra, 0)$, $(0, sb)$ and (ta, b) lying on three sides of the rectangle with diagonal from (0, 0) to (a, b). Each of r, s, t lies between 0 and 1. We may assume $t > r$, by turning the rectangle upside down if necessary. The area of the triangle is $ab(r - rs + ts)/2$. The maximum value of $r + s(t - r)$ is obtained by taking $s = 1$, $t = 1$, r arbitrary. (The case where two vertices of the triangle lie on one side of the rectangle is readily dispensed with.)

I15. If we subdivide the square into four smaller squares, each with side $\frac{1}{2}$, we see that at least one of these four squares contains 3 or more of the 9 points. Apply the preceding problem to 3 such points. (Note: it seems unlikely that $\frac{1}{8}$ is the smallest possible value in this problem, but this question is more difficult.)

I16. $\frac{1}{2}$. This can be proved by use of the corresponding result for rectangles in Problem I14. Given any parallelogram, a coordinate system can be imposed with the origin at one of the acute angles so that the vertices are

$$(0, 0), \quad (a, 0), \quad (a + b, c) \quad \text{and} \quad (b, c),$$

where a, b, and c are positive. Consider the transformation T that carries any point (x, y) into $(x - by/c, y)$. We note that T carries the parallelogram above into the rectangle with vertices

$$(0, 0), \quad (a, 0), \quad (a, c), \quad (0, c),$$

having the same area ac. Finally, it is easy to prove that the area of any triangle is invariant under the transformation T. This can be done by using the standard formula for the area of a triangle, cf. (4) in §7.1.

J1. Expanding $\Sigma (x_i - \bar{x})^2 \geq 0$ we get

$$\Sigma x_i^2 - 2\bar{x}\Sigma x_i + n\bar{x}^2 \geq 0.$$

Multiply by n and use the fact that $n\bar{x} = \Sigma x_i$.

J2. 3 tosses. The probability of getting a sum of 7 or 11 when two dice are tossed is 8/36 or 2/9. Hence the probability of getting 7 or 11 up at least once in three tosses is

$$1 - (7/9)^3 = 386/729 > ½.$$

J3. 4 tosses, with probability 13/18. The probability of *distinct* numbers on 3 tosses is (1)(5/6)(4/6) or 5/9; of distinct numbers on 4 tosses is 5/18.

J4. 5 tosses, with probability 671/1296. The probability of *not* repeating the number on the first toss is $(5/6)^{n-1}$ on n tosses of the die. Choose n just large enough so that this complementary probability is ½ or less.

J5. (i) At least two out of four is more likely iff $p > 2/3$. One way to see this is to write $q = 1 - p$, so that the question can be formulated as the inequality

$$1 - q^4 - 4q^3p > 1 - q^2.$$

Replacing p by $1 - q$ and using simple algebra, we find that this reduces to $3q^2 - 4q + 1 > 0$ or $(3q - 1)(q - 1) > 0$, which holds for $q < ⅓$. (ii) Let $M(n)$ denote the middle term in the binomial expansion of $(½ + ½)^{2n}$, so that

$$M(n) = (2n)!/[n!n!2^{2n}].$$

Then it is not difficult to prove that $M(n) > M(n + 1)$. Also we note that $P(n, 2n) = M(n) + ½[1 - M(n)]$ because $P(n, 2n)$ is the middle term of the binomial expansion plus half of all the others. It follows readily that $P(n, 2n) > P(n + 1, 2n + 2)$.

J6. 23. To prove this, consider the complementary probability that n people have *distinct* birthdays. This probability exceeds ½ if $n = 22$, but is less than ½ if $n = 23$, by the following argument. The total number of possible birthday arrangements among n persons is 365^n, because the first person may have any one of 365 birthdays, likewise the second person, likewise the third, and so on.

On the other hand, consider the number of possible birthday arrangements among n persons, if all birthdays are to be distinct. The first person may have any one of 365 birthdays, the second any one of 364 so as not to match the first, the third any one of 363, and so. Hence the total number of arrangements with distinct birthdays is

$$(365)(364)(363)\cdots(365 - n + 1),$$

there being n factors here. Dividing this by 365^n we get

$$(364)(363)\cdots(365 - n + 1)/365^{n-1}, \tag{1}$$

and this is the probability that n persons selected at random will have distinct birthdays.

With the help of a table of logarithms and a hand calculator, it is easy to establish that the expression (1) falls below ½ as n increases from $n = 22$ to $n = 23$. Or, alternatively, we can show that the reciprocal of (1) rises above 2 as n goes from $n = 22$ to $n = 23$. For $n = 22$ the calculation gives:

$$\begin{aligned}\log 365^{21} = 21 \log 365 &= 21(0.56229)\\ &= 11.80809,\end{aligned}$$

$$\begin{aligned}\log(364)(363)\cdots(344) &= \log 364 + \log 363 + \cdots + \log 344\\ &= 11.52774,\end{aligned}$$

using logarithms to base 10 to 5 decimal places. The difference here is

$$11.80809 - 11.52774 = 0.28035,$$

which is less than 0.30103, the value of log 2. The corresponding difference in the case $n = 23$ is

$$12.37038 - 12.06303 = 0.30735,$$

which exceeds log 2.

J7. Apply Theorem 10.5a to the points P, A, C, B to get $PA \cdot CB + PB \cdot AC \geq PC \cdot AB$, with equality iff P lies on the minor arc AB of the circumscribed circle of the triangle ABC. But the tirangle is equilateral, so the sides can be canceled out of the inequality.

J8. The inequality holds for every point P iff AB is the one and only largest side of the triangle. To prove this, suppose first that $AB > AC$ and $AB > BC$. Then by Theorem 10.5a we have $PA \cdot BC + PB \cdot AC \geq PC \cdot AB$ for all points P in the plane of the triangle. Dividing by AB we get

$$PA(BC/AB) + PB(AC/AB) \geq PC.$$

Since the terms in parentheses are less than 1, this implies $PA + PB > PC$. Conversely, suppose that $AB \leq AC$. Then locate P at A, and $PA + PB > PC$ is seen to be false. An analogous argument applies if $AB \leq BC$.

J9. 45°.

J10. 1 hour and 40 minutes. Denote the angle BAP by θ, as in Figure J10, where A is the location of the lighthouse. Then we see that

$$AP = 2.5 \sec \theta, \quad BP = 2.5 \tan \theta, \quad BK = 5, \quad PK = 5 - 2.5 \tan \theta.$$

The speed of travel along AP is 3 km. per hour, and it is 5 km. per hour along PK. Hence the travel time is

$$(2.5 \sec \theta)/3 + (5 - 2.5 \tan \theta)/5 = \frac{1}{2}\left[\frac{5}{3}\sec \theta - \tan \theta\right] + 1.$$

By Theorem 5.5a the expression in square brackets is minimized by taking $\sec \theta = 5/4$ and $\tan \theta = 3/4$.

J11. $c/\sqrt{3}$. To prove this suppose that the runner can reach a point $(x + c, 0)$ simultaneously with the cyclist. So the runner travels a distance $(x^2 + y^2)^{1/2}$ in the same time that the cyclist travels a distance $c + x$. Thus we have the equation

$$2(x^2 + y^2)^{1/2} = c + x,$$

because the cyclist goes twice as fast. Squaring and simplifying we get

$$4y^2 = 4c^2/3 - 3(x - c/3)^2.$$

To maximize y, we take $x = c/3$ and get $y = c/\sqrt{3}$.

J12. Let c be the width of the first canal, and k the width of the second one, on the path from A to B. Choose the point C so that $AC = c$ with the line segment AC perpendicular to the first canal, so that in moving from A to C we have "crossed" the first canal. Next choose the point D so that $CD = k$, with the line segment CD perpendicular to the second canal. The shortest path can now be described as follows: Move from A along a straight line parallel to DB until the first canal is reached; cross it with a bridge perpendicular to the canal; proceed again along a straight line path parallel to DB until the second canal is reached; cross it with a bridge perpendicular to the canal; again proceed along a straight line path parallel to DB. This last portion of the path coincides with a part of the line segment DB.

J13. (a) ½ (b) $\sqrt{3}$ (c) $\sqrt{3}$ (d) $\sqrt{6} - \sqrt{2}$. In parts (b), (c), and (d), the maximum is obtained by taking P, Q, R on the boundary. In part (c), if P, Q, R are any three distinct points on the surface of a sphere, there is a unique circle passing through them. This circle lies on the surface of the sphere, so the maximum is attained by taking P, Q, R on a great circle of the sphere, and thus the problem is the same as part (b).

J14. The answer is 1, obtained by locating six of the points at the vertices of a regular hexagon inscribed in the circle, and the seventh point at the center. To prove that no number larger than 1 is possible, divide the circle into six congruent sectors, each with an angle of 60° subtended at the center of the circle. Given any distribution of the seven points $P_1, \cdots, P_7$ inside or on the circle, at least two of these points must lie on some sector. The maximum distance between two points on such a sector is 1.

J15. 9603, from the triple 97, 99, and 100.

J16. Write β for $F - f$, so that $0 < \beta < 1$. First let the jeep make f trips from the starting point to create a fuel depot at a distance $\beta/(2f + 1)$ in the desert. Starting each of these trips with a full load, the jeep can deposit $f - 2\beta f/(2f + 1)$ loads of fuel at the depot. Then the jeep makes a final trip from the starting point with β units of fuel, arriving with $\beta - \beta/(2f + 1)$ at the depot. This gives a total of f units of fuel at the depot; now we use formula (1) to complete the solution.

J17. The analysis in the text above can be adapted to this case by introducing the point X_F coincident with S, with no change in the definitions of X_k for $k = 0, 1, \ldots, f$. The argument leading to the inequality (5) now gives $X_F X_f \leq \beta/(2f + 1)$. The right side of (6) has an additional term $X_F X_f$ in this case, and it follows that

$$SD \leq 1 + \frac{1}{3} + \frac{1}{5} + \cdots + \frac{1}{2f-1} + \frac{\beta}{2f+1}.$$

J18. $2\frac{5}{6}$ loads. We observe that $d(F) = 3/2$, so that formula (2) gives $f = 2$ because $1 + 1/3 + 1/5$ exceeds $3/2$. Thus formula (2) gives $3/2 = 1 + 1/3 + (F - f)/5$.

K1. A right circular cone of radius $2\sqrt{2/3}$ and height $4/3$; a regular tetrahedron of edge $\sqrt{8/3}$; a right circular cylinder of radius $\sqrt{2/3}$ and height $\sqrt{4/3}$; a cube of edge $\sqrt{4/3}$.

Given any inscribed cone which is not a right cone, it has smaller volume than a right cone on the same base because its height is smaller. So consider a right cone of radius r and height h. The relation $r^2 = h(2 - h)$ can be seen as follows: if A is the vertex of the cone, B the center of its base, C the center of the sphere, D the antipodal point to A on the sphere, and E any point on the rim of the base of the cone, then A, B, C, D are collinear, $\angle AED = \angle ABE = 90°$, $AB = h$ and $BD = 2 - h$. The volume V of the cone is $\pi r^2 h/3$, and so we see that

$$V = \pi h^2(2 - h)/3 = (h/2)(h/2)(2 - h)(4\pi/3).$$

Since the sum of $h/2$, $h/2$ and $2 - h$ is a constant, we maximize V by taking $h/2 = 2 - h$.

The inscribed tetrahedron is essentially the same problem. First the tetrahedron must be a *right* triangular pyramid for the same reason as in the case of the cone. Then the triangular base of the tetrahedron must be equilateral, since otherwise we could create a tetrahedron of larger volume by replacing the base by an equilateral triangle inscribed in the same circumscribed circle. The volume of the tetrahedron is $hA/3$, where h is the height and A the area of the base. But the area of the base is proportional to the area of its circumscribed circle, so the problem reduces to the case of the cone, giving $h = 4/3$ again. An easy calculation shows that the tetrahedron is regular, with edge of length $\sqrt{8/3}$.

In the case of a circular cylinder, again it must be a right cylinder. If an inscribed right cylinder has height h and radius r, then by simple geometry we see that

$$1 = r^2 + h^2/4,\ V = \pi r^2 h = \pi h(1 - h^2/4),$$
$$V^2 = (h^2/2)(1 - h^2/4)(1 - h^2/4)(2\pi^2).$$

Hence the maximum volume occurs if $h^2/2 = 1 - h^2/4$.

The inscribed rectangular solid of largest volume must have a square base for the same reason that the largest tetrahedron must have an equilateral triangle for base. Then the problem reduces at once to the case of the right circular cylinder because the area of the square base of the rectangular solid is proportional to the area of its circumscribed circle. Hence we have $h = \sqrt{4/3}$ again, and the other dimensions are the same by an easy calculation.

K2. Three; four.

K3. The points A, B, C, D with coordinates $(0, 0, 0)$, $(0, 1, 0)$, $(5, 0, 1)$, $(10, 0, 0)$ have the distance property $AD + BD > AB + AC + BC + CD$, as well as the same inequality for the squares. To prove part (ii), add the triangle inequalities for each of the four faces, for example $AC + BC > AB$ in triangle ABC. As to part (iii), the result $AC^2 + BC^2 + AD^2 + BD^2 > AB^2 + CD^2$ can be proved as follows. Take the origin of a coordinate system at A, with the x axis along AB, and with the triangle ABC in the xy plane. Thus the coordinates of A, B, C, D are expressible as

$$(0, 0, 0), (p, 0, 0), (q, r, 0), (s, t, u)$$

with $u \neq 0$ because otherwise the points are coplanar. Using the distance formula we calculate easily that the inequality to be proved amounts to

$$p^2 + q^2 + r^2 + s^2 + t^2 + u^2 - 2pq - 2ps > -2ps - 2rt,$$

after common terms are cancelled. This inequality is equivalent to

$$(p - q - s)^2 + (r + t)^2 + u^2 > 0.$$

K4. Let DP be the perpendicular from the vertex D of any tetrahedron $ABCD$ onto the plane of the triangle ABC. The point P may be inside, on the boundary, or outside this triangle. We prove that the triangles ABP, ACP, BCP, which cover the triangle ABC, have smaller areas than triangles ABD, ACD, BCD respectively. Hence we get the area inequalities

$$\Delta ABD + \Delta ACD + \Delta BCD > \Delta ABP + \Delta ACP + \Delta BCP \geq \Delta ABC.$$

To complete the proof we show that $\Delta ABD > \Delta ABP$, and the other two inequalities follow by analogy. In fact we prove that $\Delta ABP = (\Delta ABD)\cos\theta$, where θ is the angle between the planes of the two triangles. (In order for this area relationship to hold, the angle θ must be acute or 90°, not obtuse. There are two angles of intersection between intersecting planes, and θ is the smaller one.) Let PQ be the perpendicular from P to the line AB. Then DQ is also perpendicular to AB, by elementary solid geometry, and θ is the angle PQD. Since the triangle PQD has a right angle at P, we see that $PQ = DQ\cos\theta$. But DQ is the altitude from D to AB in the triangle DAB, and PQ is the altitude from P to AB in the triangle PAB. Using the base-altitude formula for the area of the triangle, we see that the area relation $\Delta ABP = (\Delta ABD)\cos\theta$ holds, so that $\Delta ABP > \Delta ABD$ since θ is not zero.

K5. The volume of the tetrahedron is one third of the area of a triangular face multiplied by the sum of the distances from P to the faces.

K6. No, for much the same reason that it is not true in general that the sum of two angles of a triangle exceeds the third.

K7. 14 by 14 by 28 inches. The problem is to maximize xyz subject to the restriction $2x + 2y + z = 84$. Since xyz can be written as $(2x)(2y)(z)/4$, the solution is given by $2x = 2y = z$.

K8. The shortest paths have lengths (i) $\sqrt{872}$, (ii) 25, and (iii) $\sqrt{1129}$. The problem can be solved by "unfolding" the room so as to flatten out a succession of wall-ceiling-floor surfaces onto a single plane. If the room is a length d, we compare three routes with the squares of the total distance travelled as follows:

(i) $(d + 6)^2 + 14^2$, via end wall, ceiling, side wall, end wall;
(ii) $(d + 10)^2$, via end wall, ceiling, other end wall;
(iii) $(d + 2)^2 + 20^2$, via end wall, ceiling, side wall, floor, end wall.

K9. $\sqrt{49 + \pi^2}$. To get this answer, we roll out the portion of the cylindrical surface extending from the eighth circle from $(4, 0, 0)$ to $(2\sqrt{2}, 2\sqrt{2}, 0)$ up to the eighth circle from $(4, 0, 7)$ to $(2\sqrt{2}, 2\sqrt{2}, 7)$. This reduces the problem to that of finding the shortest distance between two opposite vertices of a rectangle with sides of length 7 and π, because π is the length of the arc of an eighth circle of radius 4.

K10. $3\sqrt{3}$. This problem is manageable in an elementary way because a cone is a so-called "developable surface," one that can be rolled out on a plane without stretching or shrinking. (The cylinder in the preceding problem is an even more obvious example of a developable surface.) We roll out the region on the conical surface bordered by the straight line from $(0, 0, 0)$ to $(2, 0, \sqrt{5})$, the straight line from $(0, 0, 0)$ to $(-2, 0, \sqrt{5})$, and the semicircle on the cone from $(2, 0, \sqrt{5})$ to $(-2, 0, \sqrt{5})$, say the semicircle with nonnegative y coordinates. All the points on this semicircle have $z = \sqrt{5}$, and satisfy the equation $x^2 + y^2 = 4$, so the radius is 2. This region rolls out on a plane to give a sector of a circle of radius 3, because this is the distance along the cone from $(0, 0, 0)$ to $(2, 0, \sqrt{5})$. This fan-shaped sector is bounded by two radii, say CP and CQ, of length 3, and a circular arc PQ of length 2π, because this is the length of half the perimeter of the circle $x^2 + y^2 = 4$. The problem is thus reduced to finding the straight line distance PQ. To get this, we calculate the angle PCQ. This angle, subtended at the center of a circle of radius 3 by an arc of length 2π, is readily seen to be 120°. So the problem is to find the length of PQ in a triangle CPQ with $CP = CQ = 3$ and angle $PCQ = 120°$.

K11. Choose $x = ab/[a + \sqrt{c^2 + d^2}]$. This can be seen as follows. The sum $AP + PB$ equals

$$\sqrt{x^2 + a^2} + \sqrt{(x - b)^2 + c^2 + d^2}$$

by the distance formula. This sum of square roots has the same form as expression (1) in §3.5, so we can use the result (2) from that section with the proper adaptations.

K12. Six cuts. The interior subcube of the 27 requires six cuts, because no two of its faces can be cut simultaneously.

K13. $\theta = 2$ and $h = r$. It is easier to consider the equivalent problem of minimizing the surface area for a fixed volume V. The product of the three terms in the surface area S is

$$(2rh)(\theta rh)(\theta r^2) = 2\,\theta^2 r^4 h^2 = 8V^2.$$

Hence S is a minimum if $2rh = \theta rh = \theta r^2$.

K14. $2\sqrt{2}$. The square of the distance from the origin to any point (x, y, z) on the surface is

$$x^2 + y^2 + z^2 = \frac{1}{2}x^2 + \frac{1}{2}x^2 + y^2 + 64/x^4y^2.$$

The four terms in the last sum have a constant product, and hence the minimum is achieved by taking $\frac{1}{2}x^2 = y^2 = 64/x^4y^2$. This leads to four points closest to the origin, one of which is $(2, \sqrt{2}, \sqrt{2})$.

K15. Consider any n points (x_i, y_i, z_i), $i = 1, 2, \ldots, n$, on the sphere $x^2 + y^2 + z^2 = 1$. The sum of the squares of the distances between these points is

$$\Sigma\,[(x_i - x_j)^2 + (y_i - y_j)^2 + (z_i - z_j)^2]$$

where the sum is taken over all i, j satisfying $1 \leq i < j \leq n$. This can be written as

$$n\sum_1^n(x_i^2 + y_i^2 + z_i^2) - (\sum_1^n x_i)^2 - (\sum_1^n y_i)^2 - (\sum_1^n z_i)^2.$$

The first term here equals n^2 because $x_i^2 + y_i^2 + z_i^2 = 1$. The maximum can be achieved by taking the points on a unit circle, as in the solution to Problem F4.

K16. $3\sqrt{3}r$. Any three distinct points on a sphere determine a plane, which intersects the sphere in a circle. Thus the maximum is obtained by taking the points equally spaced on a great circle, using Theorem 6.3a.

K17. $2\pi r$. One way to get this maximum is with the points $(0, 0, r)$, $(r, 0, 0)$, and $(-r, 0, 0)$ on $x^2 + y^2 + z^2 = r^2$. We prove that if P, Q, and R are any distinct points on the sphere, then Arc PQ + Arc QR + Arc $RP \leq 2\pi r$. Define P_1 as the antipodal point to P on the sphere, so that PP_1 is a diameter. Then the sum of the arc lengths can be written as

$$(\text{Arc } PQ + \text{Arc } QP_1) + (\text{Arc } RP + \text{Arc } RP_1) \\ - (\text{Arc } QP_1 + \text{Arc } RP_1 - \text{Arc } QR).$$

The last expression in parentheses is positive or zero; each of the first two expressions equals πr.

L1. A circular arc with center A, with radius chosen so as to bisect the area; there are of course two analogous arcs with centers at B and C. This conclusion follows at once from the isoperimetric theorem applied to the closed path encircling A across the six triangles.

M1. $e^{1/e}$, or 1.44466..., is the largest value of x, and the limit of the sequence is e. First we prove that the sequence is increasing and bounded, and hence has a limit, if x lies in the interval I defined by $1 < x \leq e^{1/e}$. Let $g_n(x)$ denote the nth term of the sequence, and note that

$$x^{g_n(x)} = g_{n+1}(x). \tag{1}$$

Using induction we establish that if $x \in I$, then $g_{n+1}(x) > g_n(x)$ and $g_n(x) < e$. These inequalities hold if $n = 1$, and they imply

$$g_{n+2}(x) = x^{g_{n+1}(x)} > x^{g_n(x)} = g_{n+1}(x)$$

and

$$g_{n+1}(x) = x^{g_n(x)} < x^e \leq (e^{1/e})^e = e.$$

Next suppose that for some real number x the sequence converges to a finite limit y. As n tends to infinity in equation (1) above, we have $x^y = y$ and hence $x = y^{1/y}$. By Theorem 3 in the POSTSCRIPT ON CALCULUS, the largest value of $y^{1/y}$ is $e^{1/e}$, occurring iff $y = e$. Hence the largest value of x for which the sequence converges is $e^{1/e}$, as claimed. It may also be noted that for any value of x in the interval $(1, e^{1/e})$ the sequence converges to a value of y in the interval $(1, e)$, namely the smaller of the two values of y satisfying $x = y^{1/y}$. For example, in case $x = \sqrt{2}$, the two values of y are 2 and 4, and the sequence converges to 2. (If $0 < x < 1$, the convergence situation is more complicated. We refer the reader familiar with calculus to the book by Gabriel Klambauer, *Problems and Propositions in Analysis,* Marcel Dekker, New York, 1979, pp. 186–193, for a discussion of this case.)

REFERENCES

Almgren, Frederick J., Jr., and Jean E. Taylor, The geometry of soap films and soap bubbles, Scientific American, 235 (1976) 82–93.

Apostol, Tom M., Ptolemy's inequality and the chordal metric, Math. Mag., 40 (1967) 233–235.

Apostol, Tom M., Hubert E. Chrestenson, C. Stanley Ogilvy, Donald E. Richmond, and N. James Schoonmaker, Selected Papers on Calculus, Raymond W. Brink Selected Mathematical Papers, Vol. 2, Mathematical Association of America, 1968.

Ball, W. W. Rouse, and H. S. M. Coxeter, Mathematical Recreations and Essays, 12th ed., University of Toronto Press, Toronto, 1974.

Bankoff, Leon, An elementary proof of the Erdös-Mordell theorem, Amer. Math. Monthly, 65 (1958) 521.

Beckenbach, E. F., and R. Bellman, An Introduction to Inequalities, New Mathematical Library, No. 3, Mathematical Association of America, 1961.

Benson, Russell V., Euclidean Geometry and Convexity, McGraw-Hill, New York, 1966.

Berman, J., and E. Hanes, Volumes of polyhedra inscribed in the unit sphere in E^3, Math. Ann., 188 (1970) 78–84.

Bird, M. T., Maximum rectangle inscribed in a triangle, Math. Teacher, 64 (1971) 759–760.

Bliss, G. A., Calculus of Variations, Carus Monographs, No. 1, Mathematical Association of America, 1925.

Boas, R. P., Jr., A Primer of Real Functions, Carus Monographs, No. 13, Mathematical Association of America, 1960.

Boas, R. P., Jr., and M. S. Klamkin, Extrema of polynomials, Math. Mag., 50 (1977) 75–78.

Bottema, O., R. Z. Djordjevic, R. R. Janic, D. S. Mitrinovic, and P. M. Vasic, Geometric Inequalities, Wolters-Noordhoff, Groningen, The Netherlands, 1968.

Boys, C. Vernon, The Soap-Bubble, in The World of Mathematics, ed. James R. Newman, Simon & Schuster, New York, 1956, Vol. 2, pp. 891–900.

Breusch, R., Solution of Problem 4964, Amer. Math. Monthly, 69 (1962) 672–674.

Butchart, J. H., and Leo Moser, No calculus please, Scripta Math., 18 (1952) 221–226.

Chakerian, G. D., Intersection and covering properties of convex sets, Amer. Math. Monthly, 76 (1969) 753–766.

Chakerian, G. D., and L. H. Lange, Geometric extremum problems, Math. Mag., 44 (1971) 57-69.

Chong, Kong-Ming, The arithmetic mean-geometric mean inquality: A new proof, Math. Mag., 49 (1976) 87-88.

______, An inductive proof of the A.M.-G.M. inequality, Amer. Math. Monthly, 83 (1976) 369.

Coolidge, J. L., Some unsolved problems in solid geometry, Amer, Math. Monthly, 30 (1923) 174-180.

______, The lengths of curves, Amer. Math. Monthly, 60 (1953) 89-93.

Courant, Richard, Soap film experiments with minimal surfaces, Amer. Math. Monthly, 47 (1940) 167-174.

______, Differential and Integral Calculus, transl. by E. J. McShane, rev. ed., 2 vols., Nordeman, New York, 1940.

Courant, Richard, and Herbert R. Robbins, What Is Mathematics?, 4th ed., Oxford University Press, New York, 1947.

______, Plateau's Problem, in The World of Mathematics, ed. James R. Newman, Simon & Schuster, New York, 1956, Vol. 2, pp. 901-909.

Coxeter, H. S. M., Introduction to Geometry, 2nd ed., Wiley, New York, 1969.

Coxeter, H. S. M., and S. L. Greitzer, Geometry Revisited, New Mathematical Library, No. 19, Mathematical Association of America, 1967.

Curtis, C. W., Linear Algebra, 3rd ed., Allyn and Bacon, Boston, 1977.

Dantzig, George B., Maximization of a linear function of variables subject to linear inequalities, in T. C. Koopmans, Activity Analysis of Production and Allocation, Wiley, 1951, pp. 339-347.

______, Linear Programming and Extensions, Princeton University Press, Princeton, N. J., 1963.

Dantzig, G. B., and B. C. Eaves, eds., Studies in Optimization, MAA Studies in Mathematics, Vol. 10, Mathematical Association of America, 1974.

DeMar, R. F., The problem of the shortest network joining *n* points, Math. Mag., 41 (1968) 225-231.

______, A simple approach to isoperimetric problems in the plane, Math. Mag., 48 (1975) 1-11.

DeTemple, Duane, The birds and the bees, Part II, Math. Notes, Vol. 14, No. 2 (1971), Washington State University, Pullman, Wash.

Dörrie, Heinrich, 100 Great Problems of Elementary Mathematics, Dover, New York, 1965; original edition in German, 1932.

Dowker, C. H., On minimum circumscribed polygons, Bull. Amer. Math. Soc., 50 (1944) 120-122.

Draper, Norman, and Harry Smith, Applied Regression Analysis, Wiley, New York, 1966.

Dunkel, Otto, Problem 3957, Solution by E. P. Starke, Amer. Math. Monthly, 49 (1942) 64-65.

Epstein, Sheldon, and Murray Hochberg, A Talmudic approach to the area of a circle, Math. Mag., 59 (1977) 210.

Eves, Howard, A Survey of Geometry, rev. ed., Allyn and Bacon, Boston, 1972.

Ferguson, Thomas S., Mathematical Statistics, Academic Press, New York, 1967.

Fine, N. J., The jeep problem, Amer. Math. Monthly, 54 (1947) 24-31.

Fisk, Robert S., Solution of a problem on the maximum area of a certain triangle, Math. Mag., 50 (1977) 47-48.

Fletcher, T. J., Doing without calculus, Math. Gaz., 55 (1971) 4-17.

Frame, J. S., Finding extremes by algebraic means, Pentagon Mag., Fall 1948, pp. 14-18.

Franklin, J. N., The range of a fleet of aircraft, J. Soc. Indust. Appl. Math., 8 (1960) 541-548.

Fulton, Curtis M., and Sherman K. Stein, Parallelograms inscribed in convex curves, Amer. Math. Monthly, 67 (1960) 257-258.

Gaffney, Matthew P., and Lynn Arthur Steen, Annotated Bibliography of Expository Writing in the Mathematical Sciences, Mathematical Association of America, 1976.

Gale, David, The jeep once more, or jeeper by the dozen, Amer. Math. Monthly, 77 (1970) 493-501.

Gardner, Martin, On tessellating the plane with convex polygon tiles, Scientific American, July 1975, pp. 112-117; also December 1975, pp. 117-118.

Garver, R., The solution of problems in maxima and minima by algebra, Amer. Math. Monthly, 42 (1935) 435.

Gilbert, E. N., and H. O. Pollak, Steiner minimal trees, SIAM J. Appl. Math., 16 (1968) 1-29.

Graham, R. L., The largest small hexagon, J. Combin. Theory, 18 (1975) 165-170.

Greitzer, Samuel L., ed., International Mathematical Olympiads 1959-1977, New Mathematical Library, No. 27, Mathematical Association of America, 1978.

Hammer, Preston C., The centroid of a convex body, Proc. Amer. Math. Soc., 2 (1951) 522-525.

Hardy, G. H., J. E. Littlewood, and G. Pólya, Inequalities, 2nd ed., Cambridge University Press, 1952.

Heath, Thomas, A History of Greek Mathematics, Vol. 2, Oxford University Press, 1921.

Hestenes, M. R., Optimization Theory, Wiley-Interscience, New York, 1975.

Hewitt, Edwin, The role of compactness in analysis, Amer. Math. Monthly, 67 (1960) 499-516.

Hofmann, J. E., Elementare Lösung einer Minimumsaufgabe, Zeitschrift für Math., 60 (1929) 22-23.

Honsberger, Ross, Ingenuity in Mathematics, New Mathematical Library, No. 23, Mathematical Association of America, 1970, in particular Chapter 9, The isoperimetric problem, pp. 67-71.

______, Mathematical Gems I, Dolciani Mathematical Expositions, No. 1, Mathematical Association of America, 1973.

______, Mathematical Gems II, Dolciani Mathematical Expositions, No. 2, Mathematical Association of America, 1976.

______, ed., Mathematical Plums, Dolciani Mathematical Expositions, No. 4, Mathematical Association of America, 1979.

Hungarian Problem Books I and II, New Mathematical Library, Vols. 11 and 12, Mathematical Association of America, 1963.

James, Richard, A new pentagonal tiling, reported by Martin Gardner, Scientific American, December 1975, 117-118.

Jensen, J. L. W. V., Sur les fonctions convexes et les inégalités entre les valeurs moyennes, Acta Math., 80 (1906) 175-193.

Just, Erwin, and Norman Schaumberger, Two more proofs of a familiar inequality, Two-Year College Math. J., 6 (1975) 45.

Kazarinoff, N. D., D. K. Kazarinoff's inequality for tetrahedra, Michigan Math. J., 4 (1957) 99-104.

______, Analytic Inequalities, Holt, Rinehart and Winston, New York, 1964.

______, Geometric Inequalities, New Mathematical Library, No. 4, Mathematical Association of America, 1975.

Kelly, Paul, and Max Weiss, Geometry and Convexity, Wiley, New York, 1979.

Kershner, R. B., On paving the plane, Amer. Math. Monthly, 75 (1968) 839-844.

Klamkin, M. S., Inequalities concerning the arithmetic, geometric, and harmonic means, Math. Gaz., 52 (1968) 156-157.

______, Vector proofs in solid geometry, Amer. Math. Monthly, 77 (1970) 1051-1065.

Klamkin, M. S., and S. L. Greitzer, The 1976 U.S.A. Mathematical Olympiad, Math. Mag., 49 (1976) 157.

Klamkin, M. S., and D. J. Newman, The philosophy and applications of transform theory, SIAM Rev., 3 (1966) 10-36.

Klee, Victor L., A characterization of convex sets, Amer. Math. Monthly, 56 (1949) 247-249.

______, What is a convex set?, Amer. Math. Monthly, 78 (1971) 616-631.

______, Some unsolved problems in plane geometry, Math. Mag., 52 (1979) 131-145.

Kuhn, Harold W., "Steiner's" problem revisited, in Studies in Optimization, ed. George B. Dantzig and B. Curtis Eaves, MAA Studies in Mathematics, Vol. 10, 1974, pp. 52-70.

Lange, L. H., On two famous inequalities, Math. Mag., 32 (1959) 157-160; reprinted in Selected Papers on Precalculus, Reymond W. Brink Selected Mathematical Papers, Vol. 1, Mathematical Association of America, 1977, pp. 180-182.

______, Cutting certain minimal corners, Amer. Math. Monthly, 83 (1976) 361-365.

Lennes, N. J., Note on maxima and minima by algebraic methods, Amer. Math. Monthly, 17 (1910) 9.

Levenson, Morris E., Maxima and Minima, Macmillan, New York, 1967.

Mendelsohn, N. S., An application of a famous inequality, Amer. Math. Monthly, 58 (1951) 563.

Moise, E. E., Elementary Geometry from an Advanced Viewpoint, Addison-Wesley, Reading, Mass., 1963.

Mordell, L. J., Solution of Problem 3740, Amer. Math. Monthly, 44 (1937) 252-254.

Mosteller, Frederick, Fifty Challenging Problems in Probability, Addison-Wesley, Reading, Mass., 1965.

Moulton, J. Paul, Experiments leading to figures of maximum area, Math. Teacher, 68 (1975) 356-363.

Muirhead, R. F., Proofs that the arithmetic mean is greater than the geometric mean, Math. Gaz., 2 (1904) 283-287.

Nannini, Amos, Maxima and minima by elementary methods, Math. Teacher, 60 (1967) 31-32.

Neumann, B. H., On some affine invariants of closed convex regions, J. London Math. Soc., 14 (1939) 262-272.

Nitsche, Johannes C. C., Plateau's problems and their modern ramifications, Amer. Math. Monthly, 81 (1974) 945-968.

Niven, Ivan, Which is larger, π^e or e^π ?, Two-Year College Math. J., 3 (1972) 13-15.

______, Convex polygons that cannot tile the plane, Amer. Math. Monthly, 85 (1978) 785-792.

______, An ellipse problem beyond the reach of calculus, Two-Year College Math. J., 10 (1979) 162-168.

Oakley, C. O., End-point maxima and minima, Amer. Math. Monthly, 54 (1947) 407-409.

Ogilvy, C. S., Exceptional extremal problems, Amer. Math. Monthly, 67 (1960) 270-275.

______, A Calculus Notebook, Prindle, Weber & Schmidt, Boston, 1968.

Osserman, Robert, The isoperimetric inequality, Bull. Amer. Math. Soc., 84 (1978) 1182-1238.

______, Bonnesen-style isoperimetric inequalities, Amer. Math. Monthly, 86 (1979) 1-29.

Pappus (approx. A. D. 300), Isoperimetric figures, in The World of Mathematics, Vol. 1, ed. James R. Newman, Simon & Schuster, New York, 1956, pp. 207-209.

Parker, W. V., and J. E. Pryor, Polygons of greatest area inscribed in an ellipse, Amer. Math. Monthly, 51 (1944) 205-209.

Pedoe, D., A Course of Geometry, Cambridge University Press, New York, 1970.

Polachek, Harry, The structure of the honeycomb, Scripta Math., 7 (1940) 87-98.

Pólya, George, How to Solve It, Princeton University Press, Princeton, N. J., 1945.

______, Mathematics and Plausible Reasoning, Vol. 1: Induction and Analogy in Mathematics (1954); Vol. 2: Patterns of Plausible Inference, 2nd ed. (1968), Princeton University Press, Princeton, N. J., 1968.

______, The minimum fraction of the popular vote that can elect the President of the United States, Math. Teacher, 54 (1961) 130-133.

______, Mathematical Discovery, Vol. 1, Wiley, New York, 1962; Vol. 2, 1965.

______, Mathematical Methods in Science, New Mathematical Library, No. 26, Mathematical Association of America, 1977.

Pólya, G., and G. Szegö, Isoperimetric Inequalities in Mathematical Physics, Annals of Mathematical Studies, No. 27, Princeton University Press, Princeton, N. J., 1951.

Rademacher, Hans, and Otto Toeplitz, The Enjoyment of Mathematics, Princeton University Press, Princeton, N. J., 1957.

Rado, Tibor, What is the area of a surface?, Amer. Math. Monthly, 50 (1943) 139-141.

Radziszewski, C., Sur un problème extrémal relatif aux figures inscrites dans les figures convexes, C. R. Acad. Sci. Paris, 235 (1952) 771-773.

Raifaizen, Claude H., A simpler proof of Heron's formula, Math. Mag., 44 (1971) 27.

Rainwater, J., Problem 4908, Amer. Math. Monthly, 67 (1960) 479.

Reinhardt, K., Über die Zerlegung der Ebene in Polygone, dissertation, Universität Frankfurt, 1918.

Riesz, F., Sur une inégalité intégrale, J. London Math. Soc., 5 (1930) 162–168.

Rudin, Walter, Principles of Mathematical Analysis, 3rd ed., McGraw-Hill, New York, 1976.

Samuelson, Paul A., Maximum principles in analytical economics, Science, 173 (1971) 991–997, Nobel Prize lecture.

Schattschneider, Doris, Tiling the plane with congruent pentagons, Math. Mag., 51 (1978) 29–44.

Schaumberger, Norman, Some consequences of a property of the centroid of a triangle, Two-Year College Math. J., 8 (1977) 142–144.

Sokolowsky, Daniel, A note on the Fermat problem, Amer. Math. Monthly, 83 (1976) 276.

Starke, E. P., Solution of Problem 3957, Amer. Math. Monthly, 49 (1942) 64–65; proposed by Otto Dunkel, 47 (1940) 245.

Steen, Lynn Arthur, see Gaffney, Matthew P.

Stein, Sherman K., Algebraic Tiling, Amer. Math. Monthly, 81 (1974) 445–462.

Steinhaus, H., Mathematical Snapshots, Oxford University Press, New York, 1960.

Straszewicz, S., Mathematical Problems and Puzzles from the Polish Mathematical Olympiads, Pergamon Press, London, 1965.

Tierney, John A., Elementary techniques in maxima and minima, Math. Teacher, 47 (1953) 484–486.

Tietze, Heinrich, Famous Problems of Mathematics, Graylock Press, New York, 1965.

Tóth, L. Fejes, New proof of a minimum property of the regular n-gon, Amer. Math. Monthly, 54 (1947) 589.

______, An inequality concerning polyhedra, Bull. Amer. Math. Soc., 54 (1948) 139–146.

______, Regular Figures, Macmillan, New York, 1964.

______, What the bees know and what they do not know, Bull. Amer. Math. Soc., 70 (1964) 468–481.

______, Lagerungen in der Ebene auf der Kugel und in Raum, 2nd ed., Springer-Verlag, Berlin, 1972.

Valentine, Frederick A., Convex Sets, McGraw-Hill, New York, 1964.

Varner, John T., III, Comparing a^b and b^a using elementary calculus, Two-Year College Math. J., 7 (1976) 46.

Wagner, Neal R., The sofa problem, Amer. Math. Monthly, 83 (1976) 188–189.

Whyburn, Gordon T., What is a curve?, Amer. Math. Monthly, 49 (1942) 493–497; also in P. J. Hilton, ed., Studies in Modern Topology, MAA Studies in Mathematics, Vol. 5, Mathematical Association of America, 1968, pp. 23–38.

Yaglom, I. M., and V. G. Boltyanskii, Convex Figures, transl. from the Russian by Paul J. Kelly and Lewis F. Walton, Holt, Rinehart and Winston, New York, 1961.

INDEX